Germany's V-2 Rocket

Germany's
V-2
Rocket

Gregory P. Kennedy

Schiffer Military History
Atglen, PA

Dedication

This book is dedicated to the people who encouraged me—my children Pamela and James, my "little sister" Anna, and my beloved Marta.

Acknowledgments

This book represents the results of a life-long interest in the German V-2, and was produced using materials that have been accumulated over the past 30 years. Many people have helped me along the way, with my interest in the V-2, and with my professional career in the aerospace museum field.

First and foremost, it was through the mentoring of Frederick C. Durant III, retired Assistant Director for Astronautics at the National Air and Space Museum, that I found my career track. Without his early encouragement, I never would have been in a position to write this book. A special thanks is also due to Frank Winter, Robert Turner, Frederick I. Ordway III, and Mitchell R. Sharpe.

In the preparation of this particular work, Mark Santiago, George House, and Michael Smith of the New Mexico Museum of Space History graciously provided copies of photographs. Copies of reports on the American V-2 firings, particularly materials on the Hermes II flights, came from Jim Eckles at the Office of Public Affairs at White Sands Missile Range. My son, James Kennedy, helped with photo research.

The support and understanding of those around me, in particular my co-workers at the American Helicopter Museum—Andee Durborow, Nancy Phelan, Sherron Trio, Kathy Bratton, and Bob Wislowski—is greatly appreciated. Others who offered moral support, advice, and encouragement include my fellow members of the Writers' Group at the Barnes and Noble in Exton, PA.

Finally, I would like to acknowledge the help, advice, and patience of Bob Biondi and Ian Robertson at Schiffer Publishing.

Thank you, one and all.

Book Design by Ian Robertson.

Copyright © 2006 by Gregory P. Kennedy.
Library of Congress Control Number: 2005938109

All rights reserved. No part of this work may be reproduced or used in any forms or by any means – graphic, electronic or mechanical, including photocopying or information storage and retrieval systems – without written permission from the copyright holder.

Printed in China.
ISBN: 0-7643-2452-7

We are interested in hearing from authors with book ideas on related topics.

Published by Schiffer Publishing Ltd.
4880 Lower Valley Road
Atglen, PA 19310
Phone: (610) 593-1777
FAX: (610) 593-2002
E-mail: Info@schifferbooks.com.
Visit our web site at: www.schifferbooks.com
Please write for a free catalog.
This book may be purchased from the publisher.
Please include $3.95 postage.
Try your bookstore first.

In Europe, Schiffer books are distributed by:
Bushwood Books
6 Marksbury Avenue
Kew Gardens
Surrey TW9 4JF
England
Phone: 44 (0) 20 8392-8585
FAX: 44 (0) 20 8392-9876
E-mail: Info@bushwoodbooks.co.uk.
Free postage in the UK. Europe: air mail at cost.
Try your bookstore first.

Contents

	Introduction: Visionaries and Inventors	6
Chapter 1:	Early Development	8
Chapter 2:	Peenemünde, Trials and Triumphs	16
Chapter 3:	The Fi-103	23
Chapter 4:	July and August 1943: Critical Months	26
Chapter 5:	Dora	32
Chapter 6:	Field Trials and Preparations	39
Chapter 7:	V-Weapons	44
Chapter 8:	V-2, From the Inside	48
Chapter 9:	The V-2 at War	63
Chapter 10:	European Round Up	80
Chapter 11:	Hermes and White Sands	92
Chapter 12:	To the Moon	113
Appendix I:	American V-2 Firings	122
Appendix II:	The A-9/A-10 and the A-4b	125
Appendix III:	From Peenemünde to the Moon; A Chronology of Events Related to the V-2 Missile	126
Appendix IV:	The Canadian Arrow	129
	Bibliography	130
	Index	134

Visionaries and Inventors

In the history of astronautics there were three great pioneers: Konstantine Eduardovich Tsiolkovsky in Russia; Robert Hutchings Goddard in America; and Hermann Oberth in Germany. Two of the three were largely ignored in their lifetimes; the third pioneer helped inspire a chain of events that led to the world's first large liquid fuel rocket, and eventually to the Moon.

For centuries, humans looked towards the heavens and dreamed of flight among the stars. The stars, planets, and Moon seemed to beckon would-be space travelers. However, prior to the twentieth century a means of reaching celestial destinations did not exist, so the first "space flights" were the works of fiction writers. Nineteenth century author Jules Verne was one of the first modern writers to speculate about space travel. His novel *From the Earth to the Moon* proved particularly prophetic, and contained some remarkable parallels to the first lunar mission that occurred 90 years later.

Verne inspired a Russian schoolteacher named Konstantine Eduardovich Tsiolkovsky to think about space travel. Largely self-educated, Tsiolkovsky lived in the small, remote village of Kaluga. A childhood bout of scarlet fever had left him deaf and isolated from his peers. He spent his childhood and adolescent years studying. At the age of 23 Tsiolkovsky submitted several technical papers to the Society for Physics and Chemistry in Saint Petersburg that dealt with the theory of gases.

Unfortunately they duplicated previously published work, and the members of the Society intended to lambaste him for plagiarism. However, after reviewing young Tsiolkovsky's submission they realized he was not aware of any earlier work, and had developed his ideas independently. One of the members, Dimitri Mendeleyov (creator of the periodic table of the elements), wrote to him and explained that his ideas were not new, but encouraged him to continue his research. Tsiolkovsky turned his attention to another area—space travel.

Tsiolkovsky was probably the first person to seriously consider the challenges of space flight. In an unpublished manuscript written in 1883 titled *Free Space*, he described using reaction propulsion for travel in outer space. An 1895 article contained his first public mention of space travel. He also published two science fiction tales during the 1890s before turning his full attention to a more detailed analysis of rocketry and space flight.

For his 1903 work *The Rocket Into Cosmic Space*, he derived the mathematical formulae for predicting a rocket's performance, designed a manned space vehicle, and postulated on the best combination of propellants. Tsiolkovsky's space ship was teardrop shaped, with the lower two thirds housing the propellant tanks and engine. The upper section contained the crew quarters. For protection against acceleration during launch, he proposed having the crew lay in water filled tanks resembling bathtubs.

Tsiolkovsky also wrote about space stations and interplanetary travel. His works were virtually ignored when they first appeared, but were finally appreciated when Soviet rocket engineers in the 1930s recognized the genius of this deaf schoolteacher from Kaluga.

Jules Verne was not the only novelist to write about space. While Tsiolkovsky drew inspiration from Verne to lay the theoretical foundations for space travel, other science fiction writers continued to let their imaginations venture beyond the Earth. One of the most influential was Englishman H.G. Wells, who wrote *War of the Worlds* in 1897, a tale about a Martian invasion of Earth.

A young student in Massachusetts named Robert Hutchings Goddard happened to read *War of the Worlds*. After finishing the book, Goddard climbed a cherry tree on his family's farm and imagined the Martian war machines below him. This sparked a lifelong passion for rocketry, for he knew rockets were the only way to travel between planets. Goddard subsequently earned a Ph.D. in Physics from Clark University in Worcester, where he became a professor. Supported by a grant from the Smithsonian Institution, Goddard began a series of experiments with solid fuel rockets.

In 1919 the Smithsonian published Dr. Goddard's monograph "A Method of Reaching Extreme Altitudes." While most of this slim volume comprised mathematical formulae, it contained a paragraph stating it was possible to build a rocket capable of reaching the Moon. Upon striking the Moon, a charge of flash powder in the vehicle's nose would signal its arrival. The press labeled him "the loony moon man," and *The New York Times* chided him for failing to realize a rocket could not work in space because there was nothing for the exhaust to push against!

Of course, Goddard knew they were wrong. He understood Newton's Third Law of Motion, which the English scientist Sir Isaac Newton explained in the 17th century, and knew that a rocket's exhaust did not need anything to "react against" to produce thrust. In fact, Goddard performed experiments in 1915 that proved a rocket would generate thrust in a vacuum.

Despite such uninformed criticism, Goddard continued his research and launched the world's first liquid fuel rocket on March 16, 1926. Goddard's rocket burned liquid oxygen and gasoline. Launched from a cabbage patch on his Aunt Effie's farm, it reached an altitude of 41 feet and traveled 184 feet. It averaged a speed of 60 miles per hour as it arced across the sky.

Encouraged by this success, Dr. Goddard continued building rockets. Goddard's rockets grew larger and noisier until 1929, when several people called the fire department, thinking one was a crashing airplane. The Fire Marshal asked Dr. Goddard to move his experiments out of the Commonwealth of Massachusetts. Although Goddard moved his launching area to Fort Devins, Massachusetts, he realized he needed to find another locale for his rocket research.

Goddard sought a sparsely populated area with clear skies and lots of open space. A friend recommended New Mexico, and he found a suitable site outside Roswell. Supported by the Daniel and Florence Guggenheim Foundation, Goddard spent most of the 1930s developing many of the basic techniques rocket engineers take for granted today. He pioneered such techniques as gyroscopic guidance and film cooling for his rockets' combustion chambers. Sadly, Goddard's impact was not as great as it might have been, because he was very secretive about his work, and focused on obtaining patents on his developments.

The Guggenheims tried to convince Goddard to collaborate with another group they sponsored at the California Institute of Technology, under the direction of Dr. Theodore von Kármán. Goddard visited CalTech and spoke with von Kármán, but resisted exchanging data. One of von Kármán's principal assistants was Frank Malina, who managed to secure an invitation to visit Goddard at Roswell. Malina arrived at Goddard's home, called the Mescalaro Ranch, looking forward to being able to see some of his rockets, but left disappointed. While Goddard was willing to show Malina the launch tower, static test stand, and workshop, he would not let von Kármán's student see any rockets or research data.

Goddard worked with a staff of just four people. (Five, counting his wife Esther, who sewed parachutes, and was the group's photographer.) He eventually received 214 patents, including one for multi-stage rockets. Dr. Goddard offered his services to the government during both World Wars, but received meager support from the military. During World War I he developed a tube-launched rocket, a technique later adapted for the tank-killing bazooka. Goddard did not fare much better in the 1940s, when the Navy asked him to build a liquid-fuel rocket to help heavy seaplanes take off.

Ridiculed by the press and ignored by his own government, Goddard labored in relative obscurity until his death from throat cancer in August 1945. In sharp contrast to the way the United States' War Department seemed to ignore Dr. Goddard, the German *Wehrmacht* vigorously pursued rocketry. In the autumn of 1944 the world learned that Germany had developed the first large liquid-fuel rocket, which became known as the V-2 (for Vengeance Weapon-2).

Larger than anything built by Goddard, the development of the V-2 can be traced to the early 1920s, when the writings of the third great pioneer of astronautics, Professor Hermann Oberth, inspired a group of enthusiasts to create a rocket organization. This organization, and its highly public experiments, attracted the attention of German Army officers who were looking for a way around limitations on weapons construction imposed by the Treaty of Versailles.

1

Early Development

The Treaty of Versailles, which ended World War I, placed severe restrictions on the size of the German Army and the weapons it was allowed to have. Heavy artillery, chemical weapons, tanks, and aircraft were banned; the Army was limited to 100,000 men, and could only comprise 7 infantry and 3 cavalry divisions. The intent was to prevent Germany from ever being able to wage an aggressive war. However, the weapons specified were those that had been used during "The Great War." By the fall of 1930, the Army Ordnance Branch realized the Treaty made no mention of rockets, so the German Army decided to begin a rocket program.

After meeting on December 17, 1930, the Ordnance Board appointed Captain Walter Dornberger to head the nascent program. The work was to be conducted at the Kummersdorf Artillery Range, about 17 miles from Berlin. Dornberger, who had recently received his engineering degree from *Technische Hochschule Charlottenburg*, was directed to develop a liquid fuel rocket with a greater range than existing artillery. He was also directed to maintain absolute secrecy over the program. While it could be argued at this early stage the rocket program was purely experimental, and did not violate the Versailles Treaty, Article 168 specified "the manufacture of arms, munitions, or any war material, shall only be carried out in factories or works the location of which shall be communicated to and approved by the Governments of the Principal Allied and Associated Powers." Therefore it was best to hide the effort, because it could be perceived as a violation of the treaty. The rocket laboratory at Kummersdorf was named "Experimental Station West."

A survey of German newspapers showed there was a great deal of popular interest in rockets. The most widely publicized rocket experiments were those conducted by Fritz von Opel and Max Valier. German industrialist and auto manufacturer Fritz von Opel supported the spectacular, if somewhat unscientific experiments of Max Valier. During the late 1920s and early 30s, Valier and Opel applied rocket propulsion to cars, gliders, rail cars, and sleds.

Valier was one of the early members of the German *Verein für Raumschiffahrt*, or Society for Space Travel, which later became known by its initials, VfR. Professor Hermann Oberth, a mathematics and physics teacher from Transylvania, and author of *Die Rakete*

Major General Walter Dornberger headed the development of large ballistic missiles in Germany from 1930 to 1945. *Courtesy of the New Mexico Museum of Space History.*

Chapter 1: Early Development

zu den Planetenräumen (*The Rocket Into Interplanetary Space*), helped create the VfR in 1927.

Valier's experiments and notoriety did not endear him to his fellow VfR members, who discussed expelling him from the group. The newspapers often focused on Valier's more extravagant notions of how rockets might be used in the future—like trans-Atlantic passenger carrying craft, for example—which the society's members felt demeaned their more serious work. Valier was not expelled from the VfR, largely because the procedures to do so were very complicated and onerous. Also, by that time von Opel had taken the limelight, and Valier drifted into the background. Valier and von Opel parted company, and continued to work on rocket propulsion separately.

Oberth subsequently became involved in a project to build a liquid fuel rocket for the 1929 premiere of Fritz Lang's movie "The Girl in the Moon." While ostensibly sponsored by the Ufa Film Company, the effort ended up being financed by Lang and Oberth themselves. For this project Oberth hired two assistants. One of them was Rudolph Nebel, a former World War I fighter pilot with an (limited) engineering background. While Nebel had an engineering degree, he lacked any practical experience. As a youth, Nebel had read Verne (just as Tsiolkovsky had), which inspired him to become a pilot. He experimented with arming an airplane with rockets during the world war. At the time of his employment with Oberth, Nebel owned a fireworks company in Saxony.

Oberth's second assistant was an aviation student named Alexander Shershevsky, a Russian refugee and ardent Communist. Despite his political ideology, Shershevsky could not return to the Soviet Union because he had overstayed his visa, and faced prosecution in his native land. Shershevsky was always eager to engage others in political discussions, a habit that couldn't help but slow down the pace of work on the rocket.

With barely four months to design and build the rocket, they failed. Oberth tried to build a rocket intended to burn liquid oxygen and gasoline, but it never reached the flight stage, and its development took much longer than first planned. The studio publicity department had publicized the flight, and made all sorts of unrealistic claims for the expected altitude. The task was simply beyond the capabilities of Oberth and his assistants within the allotted time. When only a few weeks were left before the scheduled release of the film, a desperate Oberth even experimented with a hybrid rocket that burned liquid oxygen and a solid fuel of carbon, but he could not find a satisfactory fuel mixture.

The premiere came and went with no rocket launch. Oberth's rocket had not been completely assembled; in fact, the components were scattered among several subcontractors. A launch tower had been finished and some testing equipment gathered, but these items remained in the possession of the Ufa Company. The pieces of Oberth's rocket would remain in the hands of the various contractors until someone paid the bills for them.

In the wake of the film project debacle, Nebel proposed putting a solid fuel motor in the unfinished Oberth rocket and conducting a faked ascent for the public. After all, to the uninformed lay person, there would be no discernible difference between a liquid fuel or solid fuel rocket as it flew. He even went so far as to suggest the rocket might be rigged to explode so nobody would ever know the flight had not been made with a liquid fuel rocket. The members of the VfR quickly shouted him down. Then Nebel suggested the group build a liquid fuel Minimum Rocket—the *Minimumrakete*, or *Mirak*. Oberth opposed the idea because he wanted to build a high-altitude rocket that would show the superiority of liquid propellants over existing solid fuels. Because of its size, Nebel's *Mirak* would not perform as well as solid fuel rockets. Despite Oberth's objections, the *Mirak* idea was accepted.

The group managed to obtain the equipment from the rocket project from Ufa and the contractors, which included a liquid fuel motor Oberth called the *Kegeldüse*. The VfR demonstrated it on July 23, 1930, for the Director of the *Chemisch-Technische Reichsantalt*, or Institute for Chemistry and Technology. The motor fired for 90 seconds and produced a thrust of about 15 pounds. Oberth left Germany shortly after the successful test of the *Kegeldüse*, and returned to his teaching position as professor of mathematics at the high school of Mediash. He had been on an extended leave of absence to work on the rocket for the Lang film, and his superiors were anxious for him to return.

Johannes Winkler, who launched the first liquid fuel rocket in Germany on February 21, 1931. *Courtesy of the New Mexico Museum of Space History.*

A new VfR member helped with the *Kegeldüse* test, an engineering student named Wernher von Braun. Born on March 23, 1912, he was the second of three sons of Emmy and Magnus Freiherr von Braun. As a youth Wernher developed a keen interest in astronomy. For his confirmation, he did not receive a watch or pair of long pants like most Lutheran boys—he received a telescope. Most of the encouragement to look to the heavens came from his mother. His father, who was minister for education and agriculture for the German Republic, could not fathom where his son acquired his interest in technical matters. He was working as an apprentice in a Berlin locomotive shop when he joined the VfR.

Klaus Riedel, a young engineer, had also recently joined the VfR, and was eager to participate in the experimental program. The group set up a testing area on a farm owned by Riedel's grandparents in Saxony. Riedel's offer came at a propitious moment. Max Valier died on May 17, 1930, in an explosion at the Heylandt Company, while he was working on a rocket powered car, and there was a general clamor to ban rocket experiments altogether. The farm was remote enough that the VfR could continue testing rockets without much public attention until the uproar died down.

Initial results with the *Mirak* motor proved disappointing. At first, the *Mirak* did not produce enough thrust to be measured on the test stand the group built. After several tries the group managed to coax "a recoil of 3 or 4 pounds" from it. Then, just as they'd refined the motor to the point where it produced more thrust than its weight, it exploded.

A second *Mirak* blew up in the spring of 1931, and a third was planned. Although the *Kegeldüse* had been fired on the ground, so far nobody had ever flown a rocket using liquid propellants in Germany. They hoped one of the *Miraks* would be the first.

As things developed, the VfR did not launch the first liquid fuel rocket in Germany—Johannes Winkler did on February 21, 1931. Winkler was a past president of the VfR, but had left the organization to experiment with rockets on his own. Winkler's rocket consumed liquid oxygen and liquid methane. The rocket stood about

Test firing of a two-stick *Repulsor* at the *Raketenflugplatz*. Courtesy of the New Mexico Museum of Space History.

two feet tall and weighed 11 pounds. He launched it at the drill ground near Dessau. At the time no one in Germany knew of Goddard's success in America, so the VfR members thought it was the world's first liquid fuel rocket. Although it was not the world's first, it was the first liquid fuel rocket flown in Europe.

By the time of Winkler's rocket VfR membership had reached a thousand. The group even managed to secure a new testing location: an abandoned Army ammunition storage site in the northern suburbs of Berlin that they named the *Raketenflugplatz*, or Rocket Flying Place. The VfR leased the 300-acre facility from the City of Berlin in 1930 for the nominal sum of $4.00 per year. Ownership over the tract was somewhat murky—the City of Berlin owned the land, but the buildings belonged to the military, which insisted the group promise not to damage them.

The City would have preferred to see some sort of industrial development take place on the land, but the War Ministry's insistence that the buildings and earth works be maintained prevented that from happening. Plus, the German economy was in shambles.

Professor Hermann Oberth (left), with German inventor Hermann Ganswindt (right). Ganswindt designed a spaceship in 1891, but never carried on any actual rocket experiments. *Courtesy of the New Mexico Museum of Space History.*

Chapter 1: Early Development

Crushing war reparations imposed by the Treaty of Versailles, rampant inflation, and a worldwide depression made any large investment in developing the site unlikely. Major work would have been needed anyway, for the property was hilly and heavily forested, with marshes between the hills. Access was via a barely passable dirt road. To the small band of rocket builders, though, it was ideal.

On October 17, 1930, the VfR received the keys to the buildings on the *Raketenflugplatz*. Nebel became the "director" of the *Raketenflugplatz*. What he lacked in engineering expertise he made up with enthusiasm, as he embarked on a campaign to scrounge materials and supplies. He sent hundreds of letters soliciting equipment, tools, and raw materials from companies throughout Germany. They soon had a complete machine shop, and such materials as aluminum sheet, magnesium alloy, hardware, and tubing. The group even managed to petition for a waiver of the tax on gasoline, bringing their cost from about 80 cents a gallon to about 13 cents. Several of the buildings on the site became bachelor quarters, which enabled the VfR to obtain the services of skilled machinists by offering them free lodging. Thus, the organization continued building and launching rockets despite the chaotic economic conditions in Germany at that time.

The *Mirak*s gave way to a new series of rockets the group dubbed "Repulsors." These were modestly successful, and the VfR began to attract more attention. During their first year at the *Raketenflugplatz* the VfR launched 87 rockets, and conducted hundreds of static firings. At the same time, though, due to the worsening economic conditions in Germany, the VfR entered a period of decline. As unemployment in Germany approached 40%, many people could no longer afford the organization's annual dues of 8 Reichsmarks (RM). Membership dropped below 300, and financial support began to dry up. One of their major contributors, Hugo Hückel, canceled his pledge to the VfR, preferring to support Winkler instead. Hückel had pledged about $150.00 per month, with the condition that it be used only for experimental work, not administrative or overhead expenses. Matters were becoming critical for the VfR when the Army entered the picture.

Since being appointed to head the Army's rocket program, Dornberger tried to interest German industry in the idea, without results. After a year and a half of effort, Dornberger and his superiors realized if the German Army wanted a large rocket, they would have to develop it themselves. In the spring of 1932 Captain Dornberger, his immediate supervisor Major Ritter von Horstig, and Colonel Karl Becker, who headed the Ballistics and Munitions Branch of the Ordnance Department, visited *Raketenflugplatz*. They saw several rockets, but when the officers asked for technical data on the performance of the motors they were told none existed. Up to that time the VfR had focused on building a working rocket they could use for fund raising, rather than conducting a systematic series of scientific tests.

Acting on his own accord several months earlier, Nebel submitted a "Confidential Memo on Long Range Rocket Artillery" to Colonel Becker at the Ballistics Branch. When received, it was first dismissed as an amateur work and filed away. Becker resurrected Nebel's memo, and arranged to have him conduct a demonstration flight for the Army. In July Nebel, Riedel, and von Braun visited Kummersdorf and launched one of the Repulsors. The Repulsor only reached an altitude of 200 feet, and convinced the Army officials that rockets were, at that time, unpredictable devices at best. They did, however, show sufficient promise to warrant further work, so Dornberger took the first steps towards a large-scale rocket program. Following this visit, Dornberger hired the most promising member of the VfR, Wernher von Braun.

On November 1, 1932, von Braun reported for work at Experimental Station West. Dornberger had arranged for him to continue his studies at the University of Berlin, where he worked on a doctorate in Physics. Von Braun's first assignment was to build a 650-pound thrust rocket motor that burned liquid oxygen and a mixture of 75% ethyl alcohol and 25% water. Dornberger also hired Walter Riedel, an engineer from the Heylandt Company who had worked with Valier on his rocket powered cars, and a mechanic named Heinrich Grünow

Rudolf Nebel, "Director" of the *Raketenflugplatz* (left), and Wernher von Braun (right) at the VfR experimental center near Berlin. *Courtesy of the New Mexico Museum of Space History.*

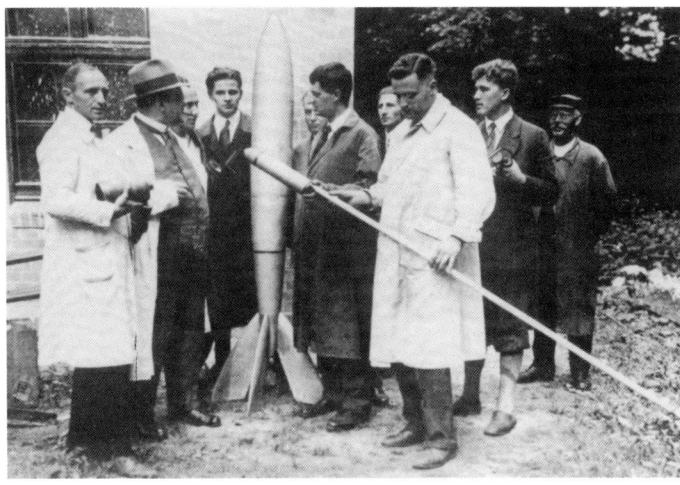

Group shot at the *Raketenflugplatz*. Rudolf Nebel is second from left; Hermann Oberth is to the right of the vertical rocket, and Wernher von Braun is second from right. *Courtesy of the New Mexico Museum of Space History.*

During the course of the Repulsor series the VfR switched from gasoline to alcohol for fuel, despite the fact that the former had a slightly higher energy content. Alcohol was cheaper, and it could be diluted with water to reduce the combustion temperature with only a minimal loss in performance. Also, it took three and a half pounds of liquid oxygen to completely burn one pound of gasoline; a pound of alcohol only needed two pounds of oxygen. These advantages offset the loss in performance by using a slightly less volatile fuel.

Drawing on his experience with the *Kegeldüse*, *Mirak*, and Repulsor rockets, von Braun had a motor ready by mid-December. On the night of December 21, 1932, von Braun tested the new motor. The first *Mirak* used a version of Oberth's *Kegeldüse* motor. The *Kegeldüse* was made from steel, with a copper lining in the combustion chamber. During static firings, the group observed the combustion chambers quickly burned through. For *Mirak* Number Two they tried placing the combustion chamber inside the liquid oxygen tank, but this too exploded. The later Repulsors had a jacket of water around the combustion chamber, but obviously for a rocket that burned for more than a few seconds this would not work. Von Braun's first engine for the Wehrmacht used a technique called regenerative cooling. The combustion chamber had an inner and outer wall. Alcohol fuel circulated through the space between the walls before entering the combustion chamber, cooling the motor.

Everyone was proud of the motor and the test stand, which had just been completed a few days before. Concrete walls, 18 feet long by 12 feet high, surrounded the test stand on three sides. The fourth side of the building comprised folding metal doors. A sliding wooden roof topped the structure. In early tests at the *Raketenflugplatz*, the data collected comprised only thrust and burn time. This new test stand would permit the collection of such data as propellant consumption rates and combustion chamber pressure. Dornberger stood, shivering, behind a fir tree a mere 10 yards from the test stand. The tree was only four inches in diameter.

Peering around his meager protection, Dornberger saw the 20-inch long pear-shaped aluminum motor gleaming in the glare from a pair of searchlights. Wernher von Braun was even closer to the motor than Dornberger. He held a can of gasoline on the end of a 12-foot pole and stood in the open end of the shed. Grünow and Riedel had the safest positions that night: behind the concrete wall monitoring propellant pressures and operating valves. When Riedel called out from behind the wall that everything was ready, von Braun lit the gasoline and held the can under the end of the motor. Riedel opened the propellant valves. A clear liquid—alcohol—dripped from the nozzle. Suddenly there was a hiss, followed by a large explosion. Shrapnel hit the searchlights, knocking them out. Alcohol and liquid oxygen dripped onto the floor, and formed puddles that burned erratically. The acrid smell of burning rubber filled the air.

Riedel and Grünow ran from behind the wall to see if anyone was hurt. At first, Dornberger and von Braun simply stared at each other in the dim light from the burning propellants, unable to believe neither was injured. Then they both began to laugh. Shards of metal were imbedded in the tree Dornberger stood behind. Fortunately the only casualties had been the motor and test stand, which was completely wrecked.

Several months later, von Braun successfully tested a 650-pound thrust motor on a rebuilt test stand. This motor proved reliable enough to place it in the *Wehrmacht's* first attempt at a flying rocket, which they dubbed the *Aggregate* (Assembly)-1, or A-1. The A-1 was 4.6 feet tall, 1 foot in diameter, and held 85 pounds of propellants. For stability, the A-1 carried a 70-pound gyroscope in its nose. This brute force approach at stability was deemed adequate for the small rocket. During a static firing the motor exploded. Rather than rebuild the shattered A-1, von Braun opted for an improved version, the A-2.

The A-2 was the same size as the A-1, and used the same motor. The difference between the two rockets was the movement of the gyroscope from the nose to the middle of the body, between the propellant tanks. This put the gyroscope near the rocket's center of gravity, where it would more effectively stabilize the craft. Two A-2s were built. Nicknamed "Max" and "Moritz," after the characters in the cartoon strip that became known as the "*Katzenjammer Kids*" in America, they flew in December 1934. The rockets reached altitudes of 1.5 miles. Even these modest rockets were too large for flight at Kummersdorf. They were launched from the island of Borkum, in the North Sea.

While von Braun built the A-1 and A-2 rockets the VfR went out of business. Nebel and Klaus Riedel kept going after von Braun left, but the organization's days were numbered. Berlin authorities were already looking askance at the rocket experiments. One of the early Repulsors crashed onto a shed owned by the Berlin Police and set it on fire. After this episode, which was recorded by the Ufa film company for one of their newsreels, the police first insisted rocket launches stop. The VfR managed to negotiate an agreement where flights could continue provided they limit the size of their

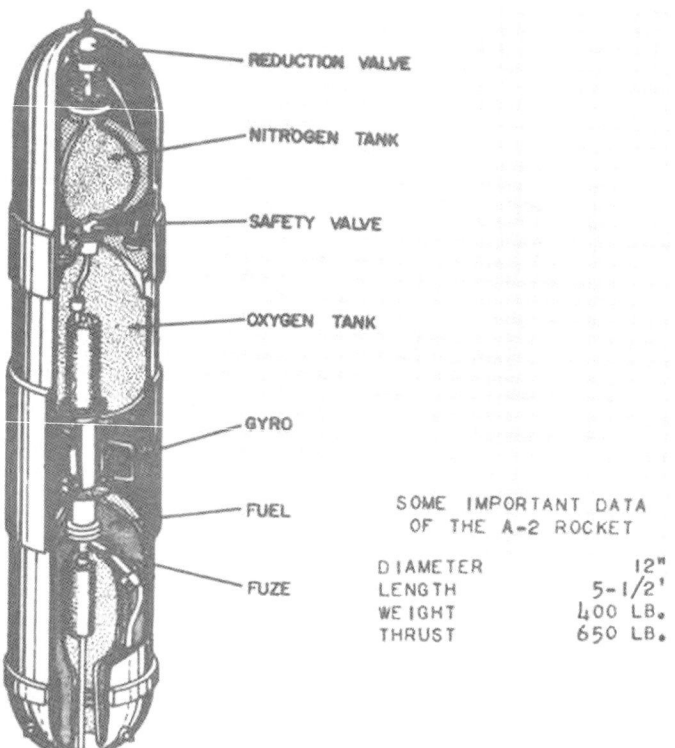

The A-2. Two rockets like this flew in December 1934.

Chapter 1: Early Development

rockets, and only fly engines that had been successfully tested on the ground three times. The death knell for the VfR came from the water department during the summer of 1934. Several faucets continually dripped at the *Raketenflugplatz*, and eventually the VfR received a large water bill—reportedly 1,600 Reichsmarks (RM). Unable to pay the bill, the VfR closed the *Raketenflugplatz*.

With the success of the A-2s the Army increased funding for the rocket program, and von Braun began work on a larger vehicle, the A-3. Dornberger continued to add to the staff at Experimental Station West, hiring many former VfR members. By the end of 1934 there were 80 people working on the rocket program at Kummersdorf.

Major Wolfram von Richtofen, a cousin of the famed "Red Baron" World War I pilot, visited Kummersdorf in January 1935. An officer in the recently formed *Luftwaffe*, and head of the Development Division of the Air Ministry, von Richtofen asked the staff at the Army Experimental Station West to investigate the possibility of putting a liquid fuel rocket motor in an airplane. The Army did not raise any objections, so by mid-summer they installed one of the 650-pound thrust motors in a Heinkel 112. Von Richtofen was so impressed with the quick response to his request that he asked the group to develop a rocket-assisted take off unit for heavy bombers. When told that Kummersdorf was too cramped to undertake such a project, he offered 5,000,000 RM for the construction of a larger facility.

This breach of protocol on the part of the junior service upset Colonel Becker, who topped the *Luftwaffe* offer by allocating 6,000,000 RM for rocket work. Prior to this, the annual budget had never been larger than 80,000 RM—suddenly the group had 11,000,000 RM to dispose of.

Everyone realized Kummersdorf was too small for flight testing large missiles, so a search began for an appropriate site. The obvious choice was to find a coastal locale that would take the flights out over water. North Sea islands, such as Borkum, where they launched the A-2s, were too close to foreign borders. Rügen, a picturesque island on the Baltic, was much better. Rockets could be flown for several hundred kilometers over the Baltic, and construction could easily be hidden among the island's dense forests, but the German Labor Front claimed it for a resort under their "Strength Through Joy" program.

While visiting his family for Christmas, von Braun's mother mentioned Usedom, near the small fishing village of Peenemünde, where his father used to go duck hunting. Usedom is one of two islands separating the Bay of Stettin and the Baltic Sea. The Peene river flows along the western side of the island. Dornberger agreed that Usedom would be a good spot, so he sought a meeting with *Luftwaffe* General Kesserling in April 1936. The evening following the meeting, Dornberger received a phone call telling him the Air Ministry purchased the necessary property from the City of Wolgast for 750,000 RM. Military construction crews soon began work on the *Heers Versuchsstelle Peenemünde* (Army Experimental Station Peenemünde), which was abbreviated HVP. The buildings were of a modern design preferred by the Air Force, rather than the gothic architecture generally favored by the Army.

Officially, HVP was a joint-services installation. The Army portion was in the forested area east of Lake Koplin; the *Luftwaffe* occupied the flatter western and northern portions of the island. The Army side was known as Peenemünde East, while the *Luftwaffe* station was Peenemünde West. The HVP was part of the *Waffenamt Prüfwesen*, or Weapons Proof and Development Office. Dornberger headed Wa Prüf 10 and 11, which were charged with solid and liquid fuel rocket development.

Development of the A-3 continued at Kummersdorf. This rocket was much larger than the A-2. It stood more than 21 feet tall, and weighed 1,650 pounds, with a motor that generated a thrust of 3,300 pounds. The A-3 was far more sophisticated than its predecessor, especially regarding the guidance system. The A-3 had a three-axis stabilization system with integrating accelerometers that would cut the engine once a pre-determined velocity had been reached. Gyroscopes in the nose were electronically linked to molybdenum vanes in the motor's exhaust. Movement of the vanes deflected the exhaust jet, and altered the course of the rocket.

An ex-Austrian Naval officer named Boykow designed the gyroscopic control system for the A-3. Regarded as the foremost expert in the area of gyroscopes, he had an interesting, if somewhat

An A-3 in a static test stand at Kummersdorf. Although the propulsion system worked well, the guidance systems did not, and all four A-3 rockets launched malfunctioned and tumbled out of control.

varied, background. Prior to 1914 he had resigned from the Navy to pursue an acting career. At the time of the A-3 project he was a director of *Kreiselgeräte G.m.b.H.* near Berlin.

During static firings at Kummersdorf everything worked fine. When deflected to simulate movement of the rocket in flight, the gyroscopes sent corrective signals to the jet vanes. In addition to the changes in the guidance system, the A-3 incorporated improvements in the propulsion system. The A-1 and A-2 rockets relied on relatively heavy bottles of pressurized nitrogen gas to force propellants from the tanks and into the combustion chamber. The A-3 had a lighter system that used liquid nitrogen. As a research vehicle, the A-3 carried a barograph and thermograph, both of which were filmed by a motion picture camera during flight. It also had instruments to measure combustion chamber pressure and skin temperature. The rocket was parachute recovered.

A small island near Peenemünde, the Griefswalder Oie, was selected as the site for A-3 launches. A feeling of optimism prevailed, as four rockets were prepared for flight. Dr. Rudolf Hermann validated the rocket's aerodynamic design in the supersonic wind tunnel at Aachen, and the 3,300-pound thrust motor performed flawlessly during static firings.

The first rocket was ready on December 3, 1937. A group of *Wehrmacht* dignitaries observed from a boat a safe distance from the island. Mechanical problems delayed the launch for several hours, much to the distress of several of the observers. Finally, the rocket lifted off. Ignition was perfect, and the A-3 smoothly climbed out of the launch tower. Then, about five seconds after launch the rocket veered to one side. The parachute ejected and was quickly burned up in the motor exhaust. By this time the rocket was tumbling, and ended its out of control flight in the waters of the Baltic.

Everyone surmised a premature parachute ejection caused the failure, so the recovery system was removed from A-3 #2. The second rocket also failed. Dense fog closed in after the second flight, so it was several days before rockets three and four could be launched. They failed, too.

Subsequent analysis showed the jet vanes provided insufficient corrective force to maintain a straight flight in winds greater than 12 feet per second. Not that this mattered very much, because the vanes responded too slowly to changes in attitude anyway, which explained why rockets three and four failed, despite being launched in calm weather. Design work was already underway for the large artillery missile (A-4), but it was obvious a great deal of research was needed before it could be built.

In April 1936, more than a year before the first flight attempts with the A-3, Dornberger set down the parameters for the A-4. Up to that time the ultimate field artillery weapon had been the Paris Gun of World War I, which could lob a shell 80 miles. As a field weapon, the Paris Gun had many drawbacks. It was very large and unwieldy, and the 210-mm diameter shell only held 23 pounds of explosive. It was also very inaccurate, with a high rate of dispersion in where the shells landed.

For his rocket weapon, Dornberger dictated that the range of the Paris Gun should be doubled, and it had to carry a one-ton warhead. Dornberger went on to impose other conditions, as well. The largest size that could be transported through railroad tunnels and small villages determined the length and fin-span of the A-4. Dispersion—that is, the circle where 50% of the rounds would be expected to land—could be no more than three-tenths of a percent of the range. For conventional artillery, the generally accepted figure was four to five percent. Thus, the A-4 had to have twice the range and one hundred times the warhead of the Paris Gun, with an order of magnitude improvement in dispersion of previous artillery.

Within a few months the preliminary design of the A-4 was completed. It would be about 45 feet long, and just over five feet in diameter, with a fin span of 11 feet. The motor would have to generate a thrust of 25 tons for 60 seconds. Propellants were liquid oxygen and alcohol, which would result in a rocket weighing just over 12 tons at launch.

In the fall of 1936 Dr. Walter Thiel joined the team at Kummersdorf to take charge of propulsion system development. At the time, the largest rocket motor in existence was the 3,300-pound, or 1.5-ton, thrust unit used on the A-3. One of the first tasks facing Thiel was how to ensure complete burning of the propellants in the combustion chamber. The A-3 motor had a relatively

Wernher Gengelback, Walter Thiel, and Hans Hueter (left to right) in front of the first successful A-4 rocket. Dr. Thiel led the team that developed the propulsion unit for the A-4. The "Frau im Mond" (Girl in the Moon) emblem was painted on paper and affixed to the rocket. Many test rounds flown from Peenemünde carried appliqués like this. The "V4" on the emblem signifies this was the fourth rocket built.

Chapter 1: Early Development

To build a combustion chamber that could produce 25 tons of thrust, Dr. Thiel clustered 18 "burner cups" like those used for his successful 1.5-ton thrust motor.

long (about 6 feet) combustion chamber. This was due in part to having the propellants injected under pressure in relatively thick streams. Contact between the streams was sufficiently violent to mix them, but a long combustion chamber was still necessary to ensure individual droplets of propellant burned before reaching the nozzle. All too often, the burning varied in intensity along the length of the combustion chamber, creating areas of heat concentration where holes burned through the motors. This was just one of the fundamental problems Dr. Thiel had to overcome to construct a large liquid-fuel motor.

To overcome the problems due to erratic combustion, injection nozzles that atomized the propellants were developed. This approach worked quite well, and the burning became more homogeneous, which in turn reduced the incidence of burn-throughs. During the next year Thiel reduced the length of the combustion chamber for a 1.5-ton motor to about a foot. Then another problem cropped up. Improvements in combustion created higher temperatures inside the motors, and they began to burn through in the nozzles and brass propellant injectors.

Thiel solved these problems by redesigning the nozzle, and recessing the propellant injectors into a cup. This latter innovation had the added bonus of providing a mixing chamber for the alcohol and oxygen before they entered the combustion chamber, further improving combustion. With a reliable, smooth running, compact 1.5-ton motor, Dr. Thiel set his sights on larger units. This presented difficulties, because experience with the 1.5-ton motors showed propellant injectors had to be carefully designed and tuned. Hundreds of hours of testing and tuning would be needed to ensure peak performance of an individual motor. Designing and building a single injector head for a 25-ton thrust motor, particularly one capable of mass production, seemed a daunting task.

Someone suggested combining groups of the injector cups used on the 1.5-ton motors to make larger motors. Thiel put three of them together, and built a successful motor with a thrust of 4.5-tons. Following this success, he proposed clustering 18 of the injector cups to create the 25-ton motor. The arrangement worked. Dr. Thiel and his staff spent the next several years refining the design of the 25-ton motor into a reliable device that could be mass produced.

The failures of all four A-3s showed there was a great deal of fundamental work remaining before the A-4 could be built, so the large missile was put on hold while the problems were worked out.

2

Peenemünde
Trials and Triumphs

In May 1937, more than six months before the A-3 launches, personnel began moving from Kummersdorf to Peenemünde. When the first group moved to the island, the staff at Experimental Station West amounted to about 90. Test stand I, for static firing large liquid fuel motors, was not finished yet, so Dr. Thiel and his group remained at Kummersdorf.

Dr. Rudolf Hermann, who did the aerodynamic testing of the A-3, joined the staff at HVP on April 1, 1937, where the Army built a new supersonic wind tunnel. After Hermann worked on the A-3 at Aachen, Dornberger and von Braun saw the need for a similar facility at Peenemünde. It took some salesmanship for Dornberger to convince his superiors such a device was necessary, but they finally agreed. Preliminary estimates for the cost to build the supersonic wind tunnel were 300,000 RM, but Dornberger figured it would cost more than three times that amount.

Working with the physical parameters set for the A-4, namely length, diameter, fin-span, operating speed, etc., Dr. Hermann began refining the missile's shape. What emerged was very different than the shape of the A-3. The A-3 had a long, slender fuselage, with small fins that extended well below the body. By comparison, the A-4 appeared much sturdier, with shorter fins.

Original plans called for building the A-4 following the A-3, but since the latter rockets failed, another test missile was needed. Designated the A-5, this rocket used the same motor as the A-3 and was about the same size, but the airframe and guidance system were completely redesigned. The A-5 was built according to Dr. Hermann's results, so it looked like a half-size version of the planned A-4.

Since the A-3's problems had nothing to do with the rocket engine, the A-5 used the same propulsion system. The alcohol tank surrounded the combustion chamber, which was about six feet long. Pressure from the evaporation of liquid nitrogen forced the alcohol and liquid oxygen from the tanks and into the combustion chamber. The A-5s would be launched from the Griefswalder Oie, site of the A-3 failures.

Designing a new guidance system took longer than expected, so the first A-5s flew without it. These flights validated the aerody-

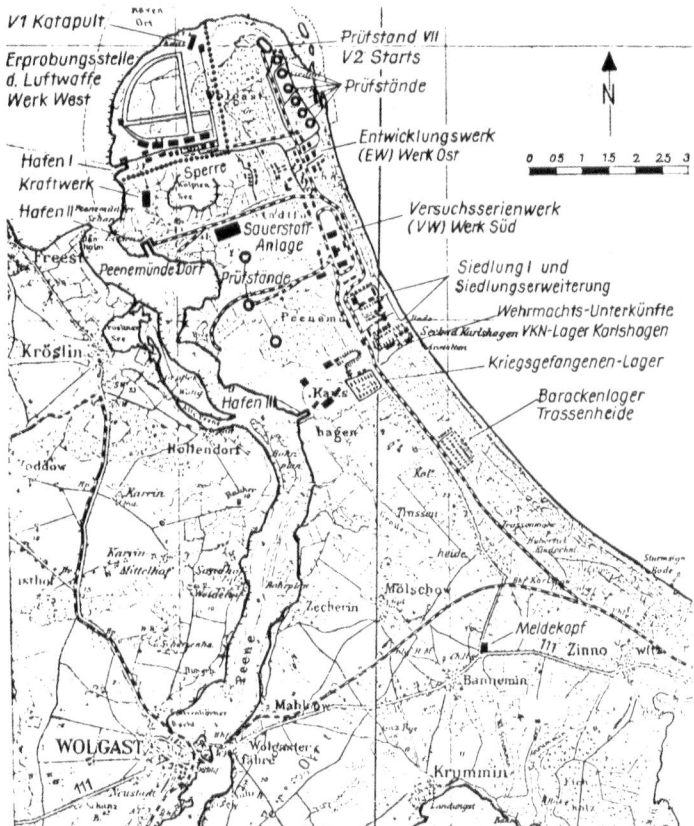

Layout of Peenemünde. A-4 firings were conducted from Test Stand (Prüfstand) VII, visible on the east side of the island. The *Luftwaffe* maintained a development station on the island's western half.

Opposite: Aerial view of Usedom Island taken in the Spring of 1944. Numerous craters are visible, particularly in the housing area. These were the results of previous Allied bombing. After the first heavy bombing raid against Peenemünde in August 1943 bomb damage was not repaired, to create the illusion the installation had been abandoned.

Chapter 2: Peenemünde, Trials and Triumphs

namic design and propulsion system. When the new guidance system was ready in the fall of 1939, the A-5 was already a flight proven vehicle. Like the A-3, the A-5 had jet vanes in the engine exhaust for steering. One important difference was that the A-5 vanes were made of carbon, rather than molybdenum, which brought the cost for a set of four vanes from 150 RM to about 1.5 RM. The carbon vanes were also stronger than the earlier ones. Other improvements included reducing the response time for the jet vanes to input from the gyroscopes. Like the A-3, the A-5 had a parachute for recovery.

Nearly two years after the A-3 rockets tumbled out of control over the Baltic, the first A-5 with the improved guidance system flew. The test was a resounding success. The first two A-5 rockets of this series flew vertically—that is, the gyroscopes held the rockets to a perfectly straight ascent—and both rockets reached altitudes of more than five miles. While these flights were certainly a significant improvement over the A-3, one important test remained.

On the third flight, the rocket flew a pre-planned trajectory, much as a combat missile would. A clockwork mechanism attached to the pitch gyroscope slowly tilted it as the rocket climbed. If the guidance system worked properly, then the vehicle's longitudinal axis would remain parallel to the axis of the gyroscope, causing the rocket to incline. After lift off the rocket climbed straight for four seconds, just as planned. Then, as the pitch gyroscope moved, the vehicle began tilting until it reached an angle of 45°. After achieving the desired angle, A-5 #3 continued to accelerate. Finally, at an altitude of 2 1/2 miles and a range of 4 miles, the parachute deployed and the rocket landed about 200 yards off shore. The new guidance system had performed perfectly.

With a successful flight of the new guidance system work proceeded on the A-4. Dr. Thiel and his staff moved from Kummersdorf in 1940 to continue their work at the recently completed Test Stand I. After that work progressed rapidly on the 25-ton thrust motor.

The A-5 continued to be used as a test vehicle for parachute and guidance system development until late 1942. Some rockets were even launched horizontally from the belly of a Heinkel 111 bomber. Thanks to the parachute recovery system, individual rockets could be flown multiple times. In all, there were about 70 flights by 25 different A-5s.

Hitler visited Kummersdorf in March 1939. During his tour he saw cutaway models of the A-3 and A-5 rockets, and witnessed static firings of several large motors. Most people who witnessed such demonstrations were awestruck. On one such visit, Hermann Göring was so impressed that he enthusiastically predicted a time when rockets would propel all sorts of vehicles, including airplanes and cars. In contrast, Hitler seemed totally unaffected. To Dornberger, Hitler's lack of response seemed strange, because the Führer was known for his enthusiasm for new weapons. Dornberger later noted that Hitler was the only visitor who did not ask any questions during the tour. Over his lunch of mixed vegetables and mineral water, Hitler discussed what he had seen with General Becker. Turning to Dornberger, he asked how long it would take to produce the A-4. The closest thing to a favorable statement made by Hitler came after lunch. As the Führer was leaving, he looked past Dornberger with a slight smile and simply said "Es war doch gewaltig," which can be translated roughly as "Well, it was grand."

In sharp contrast, the Commander in Chief of the Army, Field Marshall Walter von Brauchitsch, responded enthusiastically, and gave the rocket program the highest military priority rating in September 1939, just after the German invasion of Poland. By the spring of 1940, however, with the *Wehrmacht's blitzkrieg* victories across Western Europe, Hitler removed Peenemünde from the list of military installations authorized to receive national resources. Even worse, the center's personnel were being drafted into the military.

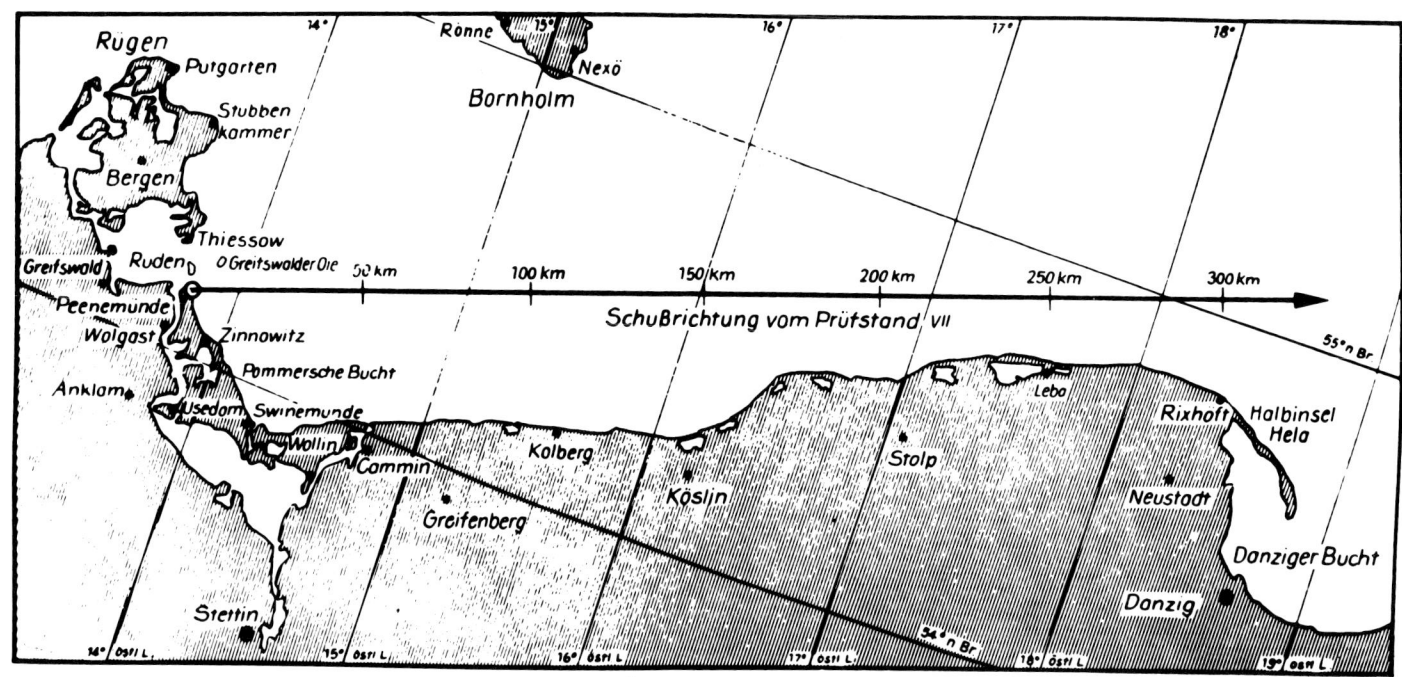

Missiles launched from Peenemünde could fly hundreds of kilometers over the Baltic Sea.

Chapter 2: Peenemünde, Trials and Triumphs

At that stage, the entire rocket program would have probably collapsed if not for the help of several high-ranking officials who found creative ways to keep it alive. The Army created the *Versuchskommando Nord* (VKN), or Experimental Command North, a combat unit assigned to the home front. Scientists, engineers, and other technical specialists who had been drafted into the Army soon found themselves reassigned to the VKN at Peenemünde.

The VKN ruse solved the personnel problems, but material shortages were another matter, and funding was barely adequate to keep the project going, with no hope of expansion. Fritz Todt, who was Minister of Armaments and Munitions, was skeptical of the rocket program, and did not fully support it. Fortunately his deputy, Albert Speer, did. Speer supervised many construction projects within the Todt organization, and authorized building facilities at Peenemünde, despite its lack of standing. Following the death of Professor Todt in a plane crash in February 1942, Speer became Minister of Armaments and Munitions.

The VKN and Speer's interest kept HVP alive, but Dornberger realized that he had to find a higher priority rating for the A-4 if he was ever going to see it turned into an operational weapon. Progress continued at Peenemünde, albeit at a slower pace than Dornberger would have wanted, and the first A-4 was finished in the spring of 1942.

A-4 #1 was used for ground handling and procedures testing. It exploded during a static firing of its motor. The first flight article was ready in June. On June 13, 1942, the first A-4 thundered off the pad. It failed. Although the motor burned for 36 seconds, the missile was unstable, rolled, and traveled less than a mile before it crashed. The next missile, launched on August 16, fared little better. It traversed 5.4 miles before the nose broke off and it exploded. After more than a decade of work and millions of Marks, some senior *Wehrmacht* officials began questioning the feasibility of using rockets as long-range weapons.

On October 3, 1942, the fourth A-4 built stood on the pad at Test Stand VII. Painted in a black and white pattern so ground cameras could record its movements in flight, the missile stood nearly 50 feet tall. It dwarfed any rockets ever built before. Everyone in the HVP organization, from Dornberger on down, knew this rocket had to succeed if the program was going to continue. Every one of the thousands of components in this rocket was inspected, checked, and rechecked several times.

Dornberger stood on the roof of the Measurements House. This was an ideal spot to observe the launch, for from here he had a nearly unobstructed view of the entire island. Colonel Leo Zanssen, the military commander of the HVP, stood next to him. "Keep your fingers crossed," Dornberger said to Zanssen, who only smiled faintly in response. Slowly, much too slowly it seemed, the countdown progressed.

Years later, Dornberger recalled how the last minutes before a launch seemed to stretch out and last much longer. "Peenemünde minutes," as he called them. In those Peenemünde minutes that day, Dornberger surveyed the installation from his vantage point. Looking through his binoculars, he spied the camouflaged Development Works. To the south was the Pre-Production Works. Even though an A-4 had not successfully flown, two large buildings had been erected for missile construction. There was the liquid oxygen plant, the power station, and the hangars of the *Luftwaffe* base on the western half of the island. There was a great deal riding on this particular missile. Dornberger saw groups of people around the island, all focused on Test Stand VII.

Wisps of liquid oxygen vapor exhausted from the missile. Vent valves let vapor from the super cold liquid escape the tank during launch preparations. Finally the vapor stopped, indicating the vent valves had been closed to build up pressure in the tank. Launch was imminent. A warning flare streaked across the sky, signaling to everyone across the island that a launch was imminent. The firing order was finally given.

"Ignition!" A shower of sparks rained from the nozzle as the propulsion engineer in the concrete blockhouse pulled the first of three levers on the control panel. The pyrotechnic igniter was burning. Seeing the sparks, he pulled the second lever, which let propellants flow into the combustion chamber. "Preliminary stage!" At this point the propellants were gravity-fed into the combustion chamber, and the motor did not produce enough thrust to lift the 13 1/2 ton rocket. Three seconds later, once he was sure the propellants were burning evenly, he pulled the third lever. This started the turbopump that force-fed liquid oxygen and alcohol into the 18 injector cups.

One of the early A-4s at Peenemünde.

Thrust quickly built up to the full 25 tons, and the rocket lifted off of the launch stand. "Rocket has lifted!" With an initial acceleration of 1-g (about 32 feet per second) the A-4 climbed slowly at first. It climbed straight up for four and a half seconds, then began a pre-planned pitch to the east. Everyone watched the rocket continue to accelerate and climb into the clear, cloudless sky. The rocket reached a pitch angle of 50°. Sixty-three seconds after lift-off *brennschluss* (thrust termination) occurred.

Its motor silent, the A-4 coasted like an artillery shell and continued to climb. The rocket reached a peak altitude of 60 miles before it began its long parabolic arc back to earth. This was the first man-made object to reach outer space. A-4 #4 impacted in the Baltic Sea 296 seconds after leaving Test Stand VII. Dr. Ernst Steinhoff, head of the Guidance Laboratory, took off in an airplane, and headed out to the predicted impact point. The impact point, as evidenced by a bright green dye marker, was 118 miles from the launch pad. The flight had been a resounding success.

"This third day of October 1942 is the first of a new era in transportation, that of space travel," declared an exuberant Dornberger at a celebration held that night.

Following this successful flight, Dornberger kept appealing for a higher priority for the missile. In the fall of 1942 he was rebuffed yet again, but he was instructed to prepare for mass production by the end of 1943, and to destroy all but one set of blueprints of the missile. This bizarre decree did not alter the A-4's low priority status, nor did it grant any additional assistance from the Ministry of Munitions. On January 8, 1943, Dornberger and von Braun both met with Speer to appeal for a higher priority for the A-4.

Speer told them that Hitler remained skeptical of the A-4, so its status would not change. However, he was placing Gerhard Degenkolb in charge of A-4 production. Degenkolb, who was known as the "Railroad Czar," had a reputation as a ruthless administrator. Dornberger tried to explain that the A-4 was vastly different and far more complex than a locomotive, but Speer expressed his confidence in Degenkolb's abilities. Speer felt Degenkolb would succeed through the sheer force of his personality. Dornberger did not share Speer's optimism, for he felt Degenkolb was incapable of understanding a device as complex as the A-4.

Immediately after his appointment Degenkolb created a production committee, and declared a production target of 900 missiles per month, starting in January 1944. At a time when most of the missiles crashed shortly after launch, and most components and subassemblies were rejected due to various defects, this was unrealistic at best.

The following month Speer unveiled a plan where the HVP would be converted into a private stock company that would be taken over by the Siemens Company. As Professor Karl Hetlage, who was in charge of financial and organizational matters in the Ministry of Munitions, explained, the government would take a cut in capital, and declare the facilities and equipment at Peenemünde were worth 1,000,000 to 2,000,000 RM. This would permit Siemens to acquire the installation at less than 1% of its cost, for the experimental station had cost more than 300,000,000 RM. If enacted, this proposal would have also put the missile program under the control of Speer's Ministry of Munitions. Fortunately for the Army, this scheme was abandoned after it became apparent that it would have disrupted the A-4 at a critical point in its development and delayed the program.

Later in the month, the Ministry of Armaments and War Production created the Long Range Bombardment Commission. Professor Waldermar Peterson, formerly one of the directors of the *Allgemeine Electrizitas Gesellschaft* (General Electric Company), chaired the Commission. This still did not raise the standing of the A-4 project, so Dornberger once again appealed to Speer.

This time, according to Dornberger, Speer told him Hitler had dreamed no A-4 would ever reach England, so the missile program's priority would not be raised. This has been questioned in recent years. For one thing, Speer did not mention the episode in his memoir *Inside the Third Reich*, and in correspondence with historian Mitchell R. Sharpe, Speer expressed doubt that even Hitler would have delayed the missile for such a reason. (Sharpe discussed the episode in *The Rocket Team*, which he co-authored with Frederick I. Ordway.) However, Dornberger insisted he saw a memorandum from Hitler stating this, which was typed in the large type-face used for the Führer. (Hitler's vanity prevented him from wearing glasses in public, and even the large-font typewriter was officially a state secret.) Whether the dream occurred or not, in the end of March, Hitler ordered the construction of a large bunker for firing missiles

Wernher von Braun (in civilian clothes) briefing a group of dignitaries at Peenemünde. *Courtesy of the New Mexico Museum of Space History.*

Chapter 2: Peenemünde, Trials and Triumphs

The launch control console in the blockhouse at Peenemünde. A periscope for viewing the launches is visible at the right.

on the French coast, near Watten, but still did nothing to raise the priority of the A-4.

Reichsführer Heinrich Himmler seemed unconcerned about the official lack of importance given to the A-4. He visited Peenemünde in early April, along with *Generaloberst* Friedrich Fromm, chief of Army Armaments, and General Emil Leeb. In his conversation with Dornberger, the *Reichsführer* expressed confidence that Hitler would soon change his mind and embrace the project. Himmler also offered to provide "protection" against sabotage and espionage. Fromm quickly pointed out that HVP was an Army installation, and the military service alone was responsible for its security.

Fromm then tried to soften the rebuff by adding that he would certainly welcome having Himmler tighten security on the rest of Usedom Island and the adjacent mainland. Himmler agreed to do this, and assigned the task to the police commissioner of nearby Stettin. Before his departure Himmler promised Dornberger that he would return without the senior Army officers to further discuss the matter.

Liftoff from Test Stand VII. The smoke streamer to the left of the rocket is a flare fired to warn of the impending launch.

Germany's V-2 Rocket

By early 1943 Himmler was actively working to create an SS (*Schutzstaffel*) industrial empire that would make the organization financially independent from the state. Himmler's plans included establishing munitions factories in the network of concentration camps he controlled. The A-4, it seemed, offered an opportunity to expand his domain.

Early A-4 Launches from Peenemünde

Missile	Date	Range	Remarks
1			Used for ground handling tests and training
2	June 13, 1942	0.8 miles	Unstable
3	August 16, 1942	5.4 miles	Nose broke off
4	October 3, 1942	118 miles	Successful flight
5	October 21, 1942	91.3 miles	Steam generator malfunctioned
6	November 9, 1942	8.7 miles	Vertical launch; reached 41.8 miles
7	November 28, 1942	5.3 miles	Tumbled, lost vanes
8			Used for crew training
9	December 12, 1942	0.06 miles	Hydrogen peroxide explosion

Source: Ley, Willy. *Rockets Missiles and Men in Space.*

The missile roars skyward.

3

The Fi-103

Not wishing to relinquish aerial bombardment to the Army, the *Luftwaffe* also worked on a long-range missile. On June 19, 1942, the German Air Ministry contracted with Gerhard Fiesler's aircraft company to build a flying bomb powered by the Argus pulse jet engine.

Pulse jets are relatively simple devices, particularly when compared to a rocket engine like the one used on the A-4. The heart of the engine is the valve assembly in the front of the engine. Flapper valves open to allow air into the engine, which is little more than a long duct. Injectors behind the valve spray fuel into the engine. A spark plug ignites the fuel/air mixture, and the resulting explosion forces the flapper valves closed. The hot gases from the explosion exhaust from the open end of the engine, which then creates a partial vacuum behind the flapper valves. The vacuum pulls the flapper valves open to admit more air, and the cycle repeats. In operation, the Argus engine repeated this cycle some 60 times per second.

The Argus engine drew upon the work of Engineer Paul Schmidt, who began working on pulse jet engines in the late 1920s. Receiving modest support through Dornberger and others during the 1930s, Schmidt actually built several pulse jet engines in the late 1930s, but none were ever flown. Schmidt designed a flying bomb in 1934, and submitted the idea to the Air Ministry, but it failed to elicit any interest at that time.

In 1939, the *Argus Moteren Gesellschaft* in Berlin also began work on a pulse jet engine. Previous to this project, Argus manufactured piston engines and superchargers for aircraft. At first the engineers at Argus were unaware of Schmidt's work, and they encountered problems with their design. Once they learned of Schmidt's engines they contacted him, and found his design for the flapper valve to be superior to theirs. During 1941 Argus incorporated Schmidt's design into what proved to be a very successful engine.

The Fiesler-103. An Argus pulse jet engine powered this cruise missile. It carried the same size warhead, and had nearly the same range as the A-4. *Source: Imperial War Museum.*

The flying bomb received the designation Fiesler 103 (Fi-103). Fiesler designed a weapon that carried a one-ton warhead over a distance of 150 miles, nearly identical to the payload and range of the A-4. In operation, though, the weapons were very different. While the A-4 would strike at supersonic speeds from the stratosphere, the Fi-103 cruised to its target at an altitude of 9,000 feet and a speed of 360 miles per hour. While the A-4 consumed propellants like liquid oxygen, alcohol, and hydrogen peroxide, the Fi-103's engine burned a low grade of fuel oil. The only exotic fuel needed was hydrogen peroxide for the flying bomb's launch catapult.

To bring the Fi-103 up to flying speed, it was launched from a catapult on a ramp. A magnetic compass kept it on the desired course. Compressed air powered the control surfaces, and forced the fuel into the engine. The missile had what looked like a small propeller on its nose. This was used to determine the range of the bomb. As it flew the propeller turned. Once the propeller had revolved a certain number of times the missile was over its target, and the elevator locked in the down position, sending the bomb into the ground.

Development of the Fi-103 took place at Peenemünde West, and the project was placed under the *Luftwaffe* Antiaircraft Artillery. As an attempted deception, the Fi-103 was called the FZG-76. FZG stood for *Flak Ziel Gerät*, or Flak Target Device. Colonel Max Wachtel commanded the 155th Flak Regiment, which was the unit organized for the Fi-103.

Construction was relatively straightforward; the Fi-103 was, after all, a pilotless aircraft. The first prototype flew on Christmas Eve 1942, a little over six months after Fiesler received the contract. It traveled 3,000 yards. Despite their relative simplicity, many of the early aircraft crashed soon after take off. The initial target date for deployment was December 15, 1943, just a year after the first flight. This proved far too optimistic.

The Fi-103 had several advantages over the A-4, the main one being that it was much cheaper to produce, which meant it could be used in greater numbers than the rocket. However, the A-4 had certain advantages. The Fi-103 cruised in a straight line at a constant speed and altitude, which made it vulnerable to both aerial interception and ground-based antiaircraft fire. The A-4 struck at supersonic speeds, giving it the advantages of surprise and immunity to all known countermeasures. Where the Fi-103 needed to be fired from prepared catapult sites, mobile field batteries could launch the A-4s.

Actually, at that time, there were two schemes for launching A-4s in the field. Most of the engineers, and Hitler himself, preferred launching from a large bunker on the French coast, where missiles could be thoroughly inspected and tested before use. Realizing how tempting a target the bunker would be for Allied bombers, military officers opted for mobile field batteries and temporary launch sites. Since they represented only fleeting targets, the field units would be difficult to hit. For the time being both options were being pursued, as Dornberger sought a compromise. There would be three firing battalions. One would be based in the large bunker, while the other two would be mobile units.

Far more fundamental, though, was the question of which weapon—flying bomb or rocket—should be fielded. The Long

A-4 test launch from Peenemünde. On May 26, 1943, the Army and *Luftwaffe* conducted firing demonstrations of the A-4 and Fi-103, respectively, for the Long Range Bombardment Commission.

Chapter 3: The Fi-103

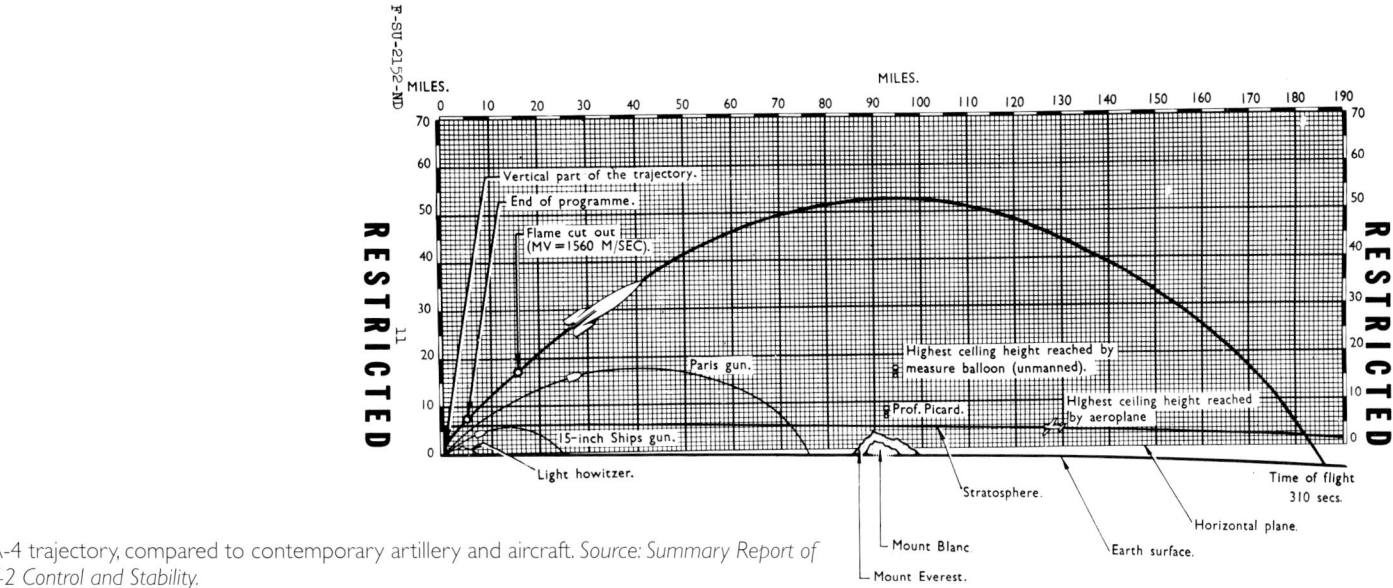

A-4 trajectory, compared to contemporary artillery and aircraft. *Source: Summary Report of V-2 Control and Stability.*

Range Bombardment Commission met at Peenemünde on May 26, 1943, to determine which weapon should be brought to production status and deployed. Commission members included Speer, *Luftwaffe* Field Marshall Erhard Milch, Grand Admiral Karl Dönitz, and Army Colonel General Fromm. After technical briefings on each weapon the group witnessed two firing demonstrations.

Both cruise missiles crashed shortly after take off, while the first A-4 traveled 165 miles. The second A-4 only reached 17 miles because the engine cut off 40 seconds after launch. These were only the 25th and 26th A-4s completed. Fortunately for both services (and the *Luftwaffe*, in particular), the visitors based their decision on the briefings, and not on the flight demonstrations. They recommended both weapons be deployed, since each had unique advantages that would complement each other.

Two days after the demonstration Speer informed Dornberger (who had already reached the rank of Colonel) that he would be promoted to *Generalmajor* effective June 1. (This was the lowest rank of General officer in the German Army during the war. Because of this, Dornberger is frequently referred to as Brigadier General in American publications, this being the equivalent rank in the U.S. Army.)

4

July and August 1943
Critical Months

The newly appointed General Dornberger, von Braun, and Steinhoff were summoned to Hitler's headquarters in East Prussia—the *Wolfsschanze* (Wolf's Lair)—on July 7, 1943. Since Steinhoff and von Braun were both pilots, the trio made the journey in a Heinkel 111 bomber that they flew. They showed Hitler models of the A-4, its proposed ground handling equipment, the bunker being built in Watten, and a film of the successful October 3 launch.

Unlike his reaction during his visit to Kummersdorf several years earlier, Hitler was thrilled after seeing the movie, and declared that had Germany possessed the missile in 1939, the current war would have been unnecessary. Hitler even apologized to Dornberger for doubting the success of the project. (He professed that Dornberger was one of only two people he'd ever apologized to in his life.) In a characteristic bombastic reaction, Hitler declared that they must produce 2,000 missiles a month, and increase the warhead to 10 tons. When Dornberger explained these were not possible, Hitler bristled, but overall the meeting went exceedingly well. Hitler awarded the A-4 the highest possible priority rating.

In an attempt to maintain security, Hitler's order that elevated the priority rating for the A-4 contained a provision that only Germans would be allowed to work on the missile. Despite such precautions, the Allies received reports on rockets, flying bombs, and other Nazi weapon projects throughout the war.

Allied intelligence had noticed unusual activity along the Baltic coast for some time. In November 1939 the British Embassy in Oslo received an anonymous report telling of German advanced weapons research, including the development of a large rocket at Peenemünde. Early in the war the report was dismissed as a German plant, because it seemed too fantastic. Only one person paid any attention to the "Oslo Report," Dr. R.V. Jones, who was in charge of tracking German weapons development for the Air Ministry's Intelligence Branch.

One of the developments described in the Oslo Report was radar. When German progress in this area came to light in 1941, this helped validate the information in the anonymous letter. Reports on the missile program at Peenemünde began trickling in, and by 1943 it seemed conclusive to the British Chiefs of Staff that something was going on there. One of the most compelling bits of information came from a pair of German prisoners of war. Unbeknownst to them, Generals von Thoma and Cruewell were being taped while they talked to each other in March 1943. General von Thoma knew about the A-4 project, and expressed his surprise to Cruewell that rocket attacks on London had not already begun. Since the pair were unaware their conversation was being monitored, von Thoma's remark was taken as genuine.

Based on the growing volume of evidence, Churchill appointed his son-in-law, Duncan Sandys, to head up a special panel specifically to investigate the German rocket program. Sandys had previously worked with solid-fuel rockets in the British Army, so he had some background in the area. Nearly everyone agreed there was a large rocket program underway in Germany—everyone, that is, except for Churchill's scientific advisor, Professor Frederick Lindemann (who became Viscount Cherwell in 1941). Lindemann just could not accept the notion that the Germans would expend the resources necessary for such a weapon, and felt the Allies were victims of an elaborate hoax.

Sir Alwyn Crow, Britain's foremost rocket expert, mistakenly theorized the Germans were developing a solid-fuel rocket weighing 100 tons, and armed with an 8-ton warhead. This was based on his work with solid-fuel rockets that burned cordite. Although he was wrong with the particulars, he at least acknowledged the existence of a German rocket program.

For some time British photographic interpreters had noted odd-looking earthworks and facilities on reconnaissance photographs of Peenemünde. In the spring of 1943 Jones received photographs showing large cylindrical objects with fins. To Dr. Jones, this offered incontrovertible proof of rockets at Peenemünde. Viscount Cherwell remained unconvinced. Interestingly, Cherwell used the fact that the rockets were painted white to make his case. He reasoned they were painted white because the Germans wanted the Allies to see them; otherwise, the missiles would have been painted in camouflage colors. Therefore, the objects were decoys meant to deceive the British.

Chapter 4: July and August 1943 - Critical Months

Field battery testing. By the middle of 1943 the level of activities at Peenemünde attracted the attention of Allied intelligence. In August 1943, the Royal Air Force staged a massive raid on the research center.

Germany's V-2 Rocket

Conflicting reports coming out of Peenemünde seemed to bolster Cherwell's arguments. The British received reports describing flying bombs launched from catapults, large rockets, and other developments. Even the photographic evidence seemed inconclusive. One day, Flight Officer Constance Babington-Smith of the Women's Auxiliary Air Force noted a small aircraft on the ground, one with swept wings and no tail. This was the rocket-propelled Messerschmitt 163 *Komet*. Dr. Jones decided there were several weapons being developed at Peenemünde. (Babington-Smith later noted yet another small aircraft, which turned out to be the Fi-103.)

Sandys sided with Jones, but Lord Cherwell stuck to his belief they were victims of a hoax. Finally, the two factions presented their arguments regarding the long-range rocket program to the Cabinet. Sandys spoke first, followed by Lord Cherwell. Jones then addressed Cherwell's objections point by point. Jones presented such a convincing case that Bomber Command was ordered to plan a massive raid on Peenemünde. There were three criteria for the mission: a full (or nearly full) moon; no clouds over the target; and clear weather over England for the bombers' return. The mission was code named Operation Hydra.

The Royal Air Force usually hit large industrial targets on moonless nights, in a technique termed "area bombing." Hydra was different. Bomber Command identified three small targets in what was a relatively small area to begin with. Success would require a high degree of precision in the bombing runs. Therefore, the raid would be conducted under a full moon. The three targets were the production area, laboratories, and housing area. The distances involved also dictated that the raid be carried out at a time when the night was long enough to permit the bombers to get away. Mid-August presented the first available time when the conditions could best be met.

Unlike most other RAF bombing raids, which were planned quickly, Hydra was on the drawing board for weeks. This would be a major raid, using all the available strength of Bomber Command. A massive raid was needed, because the RAF concluded if they failed to knock out Peenemünde with a single blow then defenses would be increased, and follow-up raids would be difficult.

The British also realized their greatest advantage would come from surprise, so they planned a tactic to confuse the Germans. Eight Mosquito fighter-bombers would stage a diversionary raid on Berlin to draw the German night fighters away from the true target. The diversionary raid was code named Operation Whitebait. The Hydra bombers were to arrive over Peenemünde 30 minutes after the diversionary raid began. Intruder units, comprising Mosquitoes and Bristol Beaufighters, also participated in the raid to attack German night fighters.

At 9:00 AM on August 17, 1943, Bomber Command issued the order to bomb Peenemünde that night. In their mission briefing, the bomber crews were told they were attacking a facility where the Germans built equipment for night air defense. Even the briefing officers remained unaware of what was going on there. The bomber crews were told that if they failed to destroy the target that night they would have to return to finish the job the very next night, so everyone wanted to make sure the night's mission succeeded. The route to Peenemünde took the heavy bombers over the North Sea, then across Denmark, before crossing the Baltic. They made their final approach to the target from over the Baltic. The bombing runs would be made from altitudes of less than 10,000 feet—6,000 to 8,000 feet was the preferred altitude. A minimum altitude of 4,000 feet was set. The total distance traveled that night would be 1,250 miles.

Luftwaffe intelligence detected an increase in radio traffic, so they knew a large bomber force was headed their way, but the British crews never once mentioned the name "Peenemünde," so the Germans were kept in the dark as to their destination. As scheduled, the Operation Whitebait aircraft dropped their flares over Berlin, which was how pathfinders marked their targets for the incoming waves of bombers. The ruse worked perfectly, and the *Luftwaffe* crews scrambled to defend the German capital. Helping the Whitebait and Hydra bombers, the *Luftwaffe* Twelfth Air Corps communication center at Arnhem-Deelen was cut off just before the diversionary raid began.

Aviatrix and test pilot Hanna Reitsch was visiting Peenemünde, and many of the senior personnel had enjoyed a reception in her honor in the officers' mess that evening. She planned to fly the *Komet* the next day. For those in attendance, *Flugkaptain* Reitsch's visit was a most pleasant end to what had been a most pleasant day.

Peenemünde existed in a seemingly cloistered environment. So far, it had remained unmolested. With its pine forests and sparkling beaches Peenemünde offered many diversions for off-duty personnel. Adding to the ambiance, it operated more like a civilian research laboratory than a military development center. Ranks were suspended in the laboratories, and what mattered most was technical expertise. Most of the civilian scientists referred to Dornberger by his civilian title, Doctor, rather than his military rank. To those German personnel stationed at the Baltic research center, the war seemed very distant.

Royal Air Force bombers frequently flew over Peenemünde on their way to other targets, so at first the approaching aircraft caused no great alarm. In the week prior to Hydra, the British deliberately sent bombers over the installation on their way to other cities; a tactic intended to lull the Germans into a false sense of complacency. The first sirens sounded around half past eleven, triggered by the Whitebait bombers heading toward Berlin. A few people headed to the shelters, but most ignored the warning. An hour later the main body of bombers approached. Again, nobody paid them much attention at first. Then the bombs began to fall.

The bombers hit Peenemünde in three waves. Each wave was to last approximately 15 minutes. The first group targeted the housing area; the next wave hit the pre-production shops; and the final group of bombers dropped their loads on the laboratory area. Peenemünde West and the test stands were not targeted. Tragically, because the Pathfinders for the first wave were off in their aim, many of the bombs hit the foreign prisoners' labor camp at Trassenheide, on the southern end of Usedom Island. Hundreds of prisoners died as a result.

The British used a recently adopted tactic for Hydra, where the mission commander circled over the target to direct the bombing runs. Group Captain John Searby was the Master Bomber for Hydra. Bomber crews also referred to him as the "Master of Ceremo-

Chapter 4: July and August 1943 - Critical Months

nies." When Searby arrived over Peenemünde he found the installation quiet. The first group of Pathfinders used radar to find the aiming point, and counted on getting a reflection from the shore of Ruden Island, a small island north of Usedom, as their way point. The radar failed to pick up Ruden, instead giving them a return from Usedom, which put their aim point two miles to the south. Searby radioed instructions to the incoming bombers to correct the aim point. Despite the instructions some of the bombers still dropped their loads on the markers and the burning camp.

West of the main base, the Germans set fire to several buildings they erected as decoys, hoping the bombers would drop their loads there. It did not work, and the RAF aircraft continued flying over the main installation.

The first wave encountered no resistance as they dropped their bombs. This was due, in large measure, to the success of the Whitebait raiders. The eight Mosquitoes that participated in Operation Whitebait each carried three 500-pound bombs, target marking flares, and bundles of "window"—strips of metal foil released to confuse enemy radar. These munitions created the illusion that Berlin was under attack.

Convinced that Berlin was the target of a major raid, the *Luftwaffe* ordered all night fighters to repel the British attack. Chaos reigned over the German capital, especially when several day-fighter units joined in the defense. Not having the proper nighttime identification lights, or being familiar with established procedures, the day fighters hindered more than they helped. Antiaircraft gunners on the ground fired at shadows, and even their own aircraft.

By the time the *Luftwaffe* pilots realized Berlin was not the true target, their aircraft were low on fuel and they had to land. Everyone converged on the nearest landing fields. About to run out of fuel, many pilots ignored ground signals telling them to land elsewhere because the runways were full. Several dozen aircraft crashed in the resulting confusion at the landing bases.

With the second wave of bombers the Germans mounted a defense. Flak batteries began firing, searchlights illuminated the sky, and ground crews started smoke generators to obscure the target. Night fighters began arriving over Peenemünde as the second wave finished dropping their bombs. Despite the smoke, the second wave dropped their bombs according to the marking flares, but Searby feared the pilots in the third wave would not be able to see them. Searby ordered the third wave to use time and distance calculations from the shore of Rügen for their bombing run. This required them to fly straight and level for a considerable distance on their final approach to the target. The third wave encountered German night fighters in force. Despite the increasing German resistance, the RAF pilots pressed their attack because of the target's importance.

Bombing from a relatively low altitude had two effects on the RAF planes. Over the target, the aircraft were buffeted from the bombs exploding so close beneath them. Their altitude also put them in range of the German light flak guns. Normally the bombers flew well above the range of the light flak batteries, but not tonight. Flak and night fighters mauled the last group of bombers, and these suffered the heaviest losses of the raid.

Despite the increasing resistance, Group Captain Searby continued to circle Peenemünde. He circled the target seven times during the raid. More than 1,500 tons of high explosive and 280 tons of incendiary bombs had been dropped on Peenemünde. Even the ratio of high explosive to incendiary bombs was different for Hydra than other raids. For a typical city target, half the bombs dropped by the RAF were incendiaries. This time, Bomber Command decided to use mostly high explosive bombs in an effort to obliterate the target.

By the time the last group departed, the crews were convinced they had destroyed the installation. Of the 596 aircraft involved in Hydra, 40 failed to return. One of the eight Mosquitoes in the Whitebait raid was shot down over Berlin, and another crash-landed in England. Counting loses among the intruders, fighters, and other aircraft that supported Hydra, the RAF lost 44 aircraft and 42 crews, which amounted to 288 men, or 6.7% of the total force. Obviously such a high loss rate could not be sustained indefinitely, but the threat posed by the A-4 justified the expenditure.

During the raid von Braun, his secretary, and several others saved technical drawings and documents from Building 4, where he had his office. When they emerged from the burning building they posted a guard armed with a bayonet over the vital documents. Dornberger rushed to the Measurements Building, which housed the guidance and control laboratories. This was one of the most important buildings at Peenemünde. He led the effort to put out the fires in the building with hand held fire extinguishers and saved the laboratories.

The next morning, Dornberger and von Braun took off in an observation plane to assess the damage. At first, the HVP looked totally devastated. Buildings lay in ruins; railroad tracks were bent or destroyed, and roads were heavily cratered. Within a few hours, though, they discovered the situation was not nearly so bad as it first appeared. Dr. Hermann's wind tunnel and the liquid oxygen plant were both unscathed. So were the docks and electrical plant on the west side of the island. All the test stands were intact. Miraculously, most of the senior staff survived. The senior-most person killed was Dr. Thiel, who died along with his family when a bomb landed directly on top of their shelter. Chief engineer Walther also died in the raid.

Altogether, there were 732 casualties at Peenemünde. More than 550 of the fatalities were among the foreign prisoners, who were locked in their barracks at Trassenheide to prevent their escape during the raid. Despite the admonition to use only German workers on the A-4, Peenemünde had a large number of forced laborers who came from occupied countries. It is estimated that the Trassenheide camp held 10,000-12,000 people.

Another large group of casualties was among the contingent of German girls from the women's auxiliary service. They had been locked in their barracks, which was standard practice to protect them from nocturnal prowlers. Their supervisor, who had the keys, died early in the raid, so the doors remained locked as the bombs fell.

Dornberger ordered that the bomb damage be left, so it would appear that Peenemünde had been abandoned. He also had crews use black and white paint on the tops of buildings so they would look burned out from the air. The ruse worked, and bombers did not hit the facility for another nine months. Actually, the facilities were quickly repaired, and the center was back in operation within six weeks.

Hydra opened an opportunity for Himmler to intervene in the A-4 project. Two days after the raid Himmler met with Hitler. During their meeting the subject of the missile program came up. Himmler had made the *Führer* aware of his interests in the A-4 for some time. His interest in the missile intensified once Hitler gave it his approval. The previous month, following the visit by Dornberger, von Braun, and Steinhoff to Hitler's headquarters, Himmler complained that the three of them should not be permitted to fly together in case the aircraft crashed. Hitler acquiesced to Himmler's suggestion, and decreed that the three were too valuable to risk in a single aircraft. This time, Himmler suggested a way to assure security of the A-4's production.

The original plans for producing the missile called for three factories: Peenemünde, the Zeppelin Works in Friederichshafen, and the Rax Works at Wiener-Neustadt, in Austria. Coincidentally, the Zeppelin Works had been hit in June, and the Austrian facility was also bombed in August. (Friederichshafen was hit because the Allies knew the Würzberg radar units were being built there.) These raids showed the impossibility of building above ground factories, so Himmler proposed moving missile production underground. As a further security measure, he also suggested using concentration camp labor to build the missiles. Since the laborers had no contact with the outside world security would be assured.

Peenemünde and the proposed production plants were not the only rocket-related facilities hit that summer. Allied intelligence could not help but notice the bunker being built near Watten, because it was so large. When finished, the bunker would have storage facilities for 108 missiles, a three-day fuel supply, and a liquid oxygen plant. Rail lines were extended to the bunker site. The project consumed 120,000 cubic meters (4,237,000 cubic feet) of concrete. Hitler's advisors pointed to the likelihood of this bunker being bombed by the Allies. He replied that every bomb that fell on it would be one less bomb to land on a German city.

Shortly after the raid on Peenemünde Allied bombers hit the bunker, just as the concrete was beginning to set. As later recalled

Pilot production plant at Peenemünde. This was one of three proposed facilities for the mass production of the A-4. British bombing in the summer of 1943 forced the Germans to move their missile production line underground.

Chapter 4: July and August 1943 - Critical Months

by General Dornberger in his memoir *V-2*, the bunker "was left a fantastic heap of wet concrete, steel, and timber." When the concrete set after the raid, the bunker was unsalvageable as a launch site, though work continued to save the liquid oxygen plant.

The desire to build a large storage and launch bunker did not end at Watten. An even more grandiose plan was concocted for a limestone quarry at Wizernes. This time, the Germans planned a novel construction scheme. A 20-foot thick concrete dome was to be erected over the quarry as the first step. Once the cover was finished, the limestone was to be excavated from underneath it, and supports added for the dome. The belief was that the dome would offer sufficient protection for the construction beneath it.

Despite the efforts devoted to building underground factories and launching bunkers, the A-4 was still nowhere near being ready for deployment as a weapon. Much work remained to train crews and perfect the missile itself. Each rocket was still being hand crafted, and included refinements and changes from the one before it. Malfunctions were frequent occurrences.

So far, ballistic data for the dispersion of the missiles was lacking. Such information would require observation of the precise impact points of the rockets and their performance. Peenemünde had a serious disadvantage for collecting such information. When launched the rockets impacted in the water, and dye markers indicated where they hit. Since the dye markers corresponded with predictions of the missiles' range, everyone assumed they functioned correctly up to impact. To observe the missiles throughout their entire flight, an overland range was needed. The *Wehrmacht* also needed a training center for operational crews. This presented another opportunity for Himmler to further involve himself in the A-4. He offered an SS training ground named the Heidelager, near Blizna, Poland, as the new launch site. The missiles would impact in the Pripet Marshes near the Bug River, almost 200 miles away.

The Army was reluctant to accept the *Reichführer's* offer to use the site in Poland at first, citing concerns over liability for damages caused by crashing rockets. After being informed that the *Reichsführer SS* would be responsible for damages outside the Heidelager, and that they need only concern themselves with safety in the immediate area, Dornberger hesitantly accepted Himmler's offer.

The remains of the proposed bunker at Watten after being bombed by the Allies. The raid occurred shortly after the Germans finished pouring concrete. According to General Dornberger, the bunker "was left a fantastic heap of wet concrete, steel, and timber."

5

Dora

Gerhard Degenkolb, who Albert Speer appointed as head of A-4 production, was one of the individuals responsible for the reorganization of the German armaments industry in 1940. Prior to the reorganization, the Armed Forces High Command and Ordnance Departments of the military services controlled all armaments production. The reorganization transferred responsibility and control of the armaments industry from the military to the civilian Ministry of Munitions. This was one of the events that reportedly led to the suicide of General Becker, a man Dornberger held in great esteem. Now Degenkolb, who Dornberger despised, was in charge of bringing the missile to production status.

Responding to pressure from his superiors, Degenkolb was eager to begin mass production of the A-4 immediately. Recent successful tests at Peenemünde had mistakenly been interpreted as a sign that the missile was ready for production. However, individual missiles were still being hand crafted in the pre-production shops at Peenemünde. Dornberger tried to point this out, but Degenkolb ignored him, and in July 1943 ordered the three factories to begin mass production.

At first, Degenkolb set a production target of 900 missiles per month. Karl Saur, chief of production and development in the Ministry of Munitions, insisted on a far more ambitious target of 2,000. Degenkolb and Saur compromised at a monthly target of 1,800 missiles.

When Himmler proposed that production take place in an underground factory, Degenkolb began searching for a suitable site. He found one in the Harz Mountains, near the town of Nordhausen. The site, which was next to the village of Niedersachswerfen, comprised two parallel tunnels with 47 cross galleries cut through a mountain. It had been excavated in 1936 for use as an oil storage area.

A state-owned corporation, Mittelwerk, was created to manufacture the A-4 missiles. Himmler provided the labor from concentration camp prisoners. With the creation of the concentration camps, the Nazi regime discovered a vast pool of cheap labor that was used by many German industries. "Wages" for the workers were paid to the SS, which operated the camps. Security for a project like the A-4 was assured, since the prisoners had no contact with the outside world.

The first group of prisoners arrived at Niedersachswerfen in August. They were housed in tents in a work camp code named "Dora." The following month,

SS Brigadeführer Hans Kammler was appointed to oversee the construction of Mittelwerk and Dora, which was a sub-camp of Buchenwald. Kammler originally trained as an architect, and previously supervised construction projects for the *Luftwaffe*. He also had been involved in the design and construction of the crematoria at Auschwitz and the destruction of the Warsaw ghetto. Speer described him as a "cold, ruthless schemer, a fanatic in the pursuit of a goal, and as carefully calculating as he was unscrupulous." Despite his high rank, he'd never served in the armed forces, and had no formal military training.

Less than a month later, the prisoners moved from the tents into the tunnels beneath Kohnstein Mountain. Living conditions in the dank chambers were unbelievably hellish. Considerable work remained to prepare the tunnels for the creation of a modern rocket factory. Most of the work had to be done by hand. The prisoners choked on the ammonia-filled dust as they swung their pick axes. Underfed, overworked, and racked by disease, many inmates died from exhaustion. Others died from the frequent beatings administered by the SS guards and prisoner-overseers, called "Kapos."

Political prisoners and others who opposed the Nazis made up most of the Dora population, but there were also a significant number of criminals imprisoned there. The Kapos most frequently came from the ranks of common criminals. As overseers, they received special privileges like above ground billeting, private rooms, and extra rations. Their status depended upon their ability to achieve results from their fellow prisoners. Rather than risk losing these few amenities, the Kapos drove their fellow prisoners to work harder, and frequently surpassed the SS in their sadistic and brutal treatment of other inmates.

Those who did not die from malnutrition, 12-hour work days, or beatings, usually died from disease. Sanitation was virtually non-existent in the tunnels, and there was no clean water. Seeing people

Chapter 5: Dora

Oberkommando des Heeres
(Chef der Heeresrüstung und Befehlshaber des Ersatzheeres)

Wa Prüf 11
Nr: DE 12 0011-5565/43

Bb.Nr. 14/11/43 g.Kdos

Firma
Mittelwerke G.m.b.H.
Berlin-Charlottenburg
Bismarckstr. 112

Ihre Zeichen

Kriegsauftrag

Es werden hiermit

K/H – R Sp II
Der Auftrag Nr 0011-5565/43 wird als Auftrag der Sonderstufe DE12 bestätigt.

Dienststempel

Berlin W 35
Tirpitzufer 72-76

19.10.43

im Auftrag gegeben:

Geheime Kommandosache
1. Dies ist ein Staatsgeheimnis im Sinne des §...
2. Nur von Hand zu Hand oder an persönliche Anschrift...
3. Beförderung möglichst durch Kurier oder Vertrauenspersonen; bei Postbeförderung unter Wertangabe von mehr als 1000 RM.
4. Vervielfältigung jeder Art sowie Herstellung von Auszügen verboten.
5. Aufbewahrung unter Verantwortung des Empfängers im Panzerschrank, ausnahmsweise im Stahlspind mit Kunstschloß.
6. Verstöße hiergegen ziehen schwerste Strafe nach sich.

Fertigungsbericht — nicht — erforderlich (Unzutreffendes streichen)
Die Angaben der nachstehenden Berichtszeile (Reichsbetriebs-Nr usw.) sind bei Weitergabe des Auftrages oder von Auftragsteilen den Unterlieferern genau und vollständig bekanntzugeben. Im Schriftverkehr genügt die Angabe der Auftrags-Nr.

Reichsbetriebs-Nr	Rü Jn	Auftraggeber (Dienststelle)	Bedarfs-Gr.	Auftrags Nr	HML	Art
	03	Oberkommando des Heeres	1431	0011-5565/43	H	1
Reichswaren-Nr						
			Auftragssteuerungs-Nr. 1431/48			

Lfd. Nr	Menge	Gegenstand (einschl. Zeichnungs-Nr, Techn. Lieferbed.-Nr, Reichswaren-Nr)	Einzelpreis je	Gesamtpreis
1.		Fertigung von 900 Geräten A4 je Monat bis zu einer Gesamtzahl von 12000 Stück ohne elektr. Inneneinrichtung, ohne Nutzlastspitze und ohne Verpackung.		
2.		Endmontage dieser 12000 Geräte einschließlich Inneneinrichtungen, Nutzlastspitze und Verpackung zum Richtpreise von: pro Stück	40000,-- RM	480000000,--

Auftragserteilung erfolgt vorbehaltlich einer nachträglichen Preisvereinbarung. Die LSÖ (Leitsätze für die Preisermittlung auf Grund der Selbstkostenbei Leistungen für öffentliche Auftraggeber vom 15.11.38) sind der Preisermittlung zu Grunde zu legen. Dementsprechend ist eine Auftragsbestätigung sofort einzusenden.

The original order for 12,000 A-4 missiles. This order was dated October 19, 1943.

Kohnstein Mountain, near Nordhausen. The A-4 production line was built beneath this mountain. *Source: U.S. Strategic Bombing Survey.*

go mad from thirst and resort to licking moisture from the rock walls was an all too familiar sight in the tunnels. Pneumonia, dysentery, tuberculosis, and typhus were rampant. A prisoner had to have a temperature of at least 104° before being admitted to the infirmary, and "medical care" usually amounted to light duty or (in extreme cases) bed rest.

Although Dora was not an extermination camp, many people died there. The prosecution brief for the Nordhausen War Crimes Trial put the figure at 20,000. At first, bodies were collected and shipped to Buchenwald for cremation. When Dora became an autonomous camp, a crematorium was built there. As the prisoners labored and died in what has been called "the hell of concentration camps," final plans took shape for A-4 production.

Liquid oxygen, which the A-4 used as an oxidizer, is very perishable, and has to be transported in insulated containers because it boils at -297°F. The ability of German industry to produce this pale blue liquid was one of the factors that determined how rapidly missiles could be launched. Despite all the production pronouncements, there was not enough liquid oxygen available to fire 2,000 missiles, or even 1,800 missiles every month. Germany could only produce enough liquid oxygen for fewer than 1,000 missiles per month. The production quota was halved to 900 per month, which was the original target set by Degenkolb. A contract was awarded to Mittelwerk on October 19, 1943, for 12,000 missiles. The total value of the contract was 480,000,000 RM, with a unit price of 40,000 RM.

The Ministry of Munitions appointed a board of directors for Mittelwerk, comprising Dr. Ing. Kurt Kettler, *SS Sturmbannführer* Otto Forschner, and Otto Bersch. A fund of 10,000,000 RM was placed at their disposal to establish the factory. The directors could draw from this fund at an interest rate of 3.25%. The initial fund proved insufficient, so the German Trust Company for War Industries allocated another 50,000,000 RM for Mittelwerk.

The main tunnels were each 5,700 feet long, 35 feet wide, and 25 feet high. For the first 800 feet from the entrances, the main tunnels were faced in concrete. The walls and ceilings were painted white to improve lighting conditions, and waterproof the rock. Cross galleries were, on average, 30 feet wide and 22 feet high. Some were deeper to accommodate large missile components.

The layout of the factory had been based on a production of 1,800 missiles per month, so when the production was halved, space could be reallocated for other projects. A-4 production took place in cross tunnels 16-41, and the adjacent main tunnels. The rest of the Mittelwerk was given over to Fi-103 and Jumo aircraft engine production.

One of the provisions of the A-4 production contract let the Mittelwerk board of directors adjust the price based upon actual costs to build the missiles. About a month after the contract was awarded, they announced the first thousand missiles would cost 100,000 RM apiece, after which the unit price would drop by 10,000 RM for every 1,000 rockets built, until it reached 50,000 RM.

Orders for component parts were placed with German industry in the late fall. To avoid bottlenecks caused by Allied bombing, the Speer Ministry ordered that there be at least two suppliers for each component. The one exception to this policy was the steam turbine in the propellant turbopump. The Ernst Heinkel Company

Chapter 5: Dora

in Jenbach, Tirol, was the only firm that could meet the specifications for this critical component. As a further safeguard against production disruptions, the supplier network was dispersed throughout Germany.

There was considerable pressure to begin production as soon as possible, even though the design was not yet frozen. Many test launches still failed, and the Peenemünde engineers were still refining and changing the design. The goal was to have the first missiles off the production line by New Years' Eve, 1943. Speer and his staff visited Mittelwerk on December 10, 1943, to check on the progress.

Gravity rollers, overhead cranes, power conveyors, and rolling jigs were all ready. The pitiful state of the slave laborers stood in sharp contrast to the modern, up to date equipment. The visitors from the Ministry of Munitions were shocked by the barbarous conditions, and some of Speer's staff "had to be forcibly sent on vacations to restore their nerves." Speer ordered construction of above ground barracks, and directed his medical staff to take steps to improve hygienic conditions at the camp. Improving the living conditions of the inmates could improve their productivity.

The first four missiles rolled off the assembly line on December 31, 1943, so it could be claimed that the target had been met. However, there were so many defects that all four had to be sent back through the factory. Serial production began in January 1944 with a production of 50 rockets. Production proved problematical, for the changes constantly introduced by the Peenemünde engineers had to be incorporated in the assembly line.

During manufacture, the missile mid section was assembled first. After being cut and formed, the sheet steel skin was spot welded to the internal longerons and ribs. A jig, similar to one that would have been used to manufacture an airplane fuselage, aligned the parts during assembly. Next, the mid section shells (which were assembled in halves) were placed in cradles to hold them while the workers installed electrical and pneumatic lines. Each mid section shell was then lined with a layer of glass wool insulation.

The tanks, which were fabricated elsewhere in the plant from welded aluminum, were installed, and the mid section shells assembled. The completed mid section traveled along the main tunnel to cross tunnels #28 and #29, where the control compartment and propulsion unit, respectively, were assembled. After these as-

One of the main tunnels, where midsections are being assembled with propellant tanks and body shells. The containers in the foreground hold rock wool insulation.

semblies were added, the missile continued along the assembly line to cross tunnel #37 for the tail unit. The completed rockets were checked in tunnel #41, which had a 50-foot ceiling so they could be inspected in a vertical position.

Warheads were not manufactured at Mittelwerk. They were loaded on the delivery trains enroute to the firing sites, where they were attached to the rockets just before launch.

Following the vertical checkout in tunnel #41, the rockets were lowered to a horizontal position and placed on rail cars for shipment to storage dumps near the launch sites. When production began at Mittelwerk, it took 15,000 man hours to build an individual rocket. Once the factory began operating at full capacity, this figure dropped to 8,000 man hours.

Dr. Georg Rickhey became general manager of Mittelwerk in April 1944. In this capacity, he reported directly to Professor Karl Hetlage of the Ministry of Munitions. Hetlage bore sole responsibility for the A-4 within the Ministry's financial and organizational planning section. Rickhey was over the previously appointed board of directors. Kettler and Bersch became directors on Rickhey's staff; Forschner was transferred to another post.

Six months after production began, more than 8,000 people worked on the assembly line. Five thousand of these were slave laborers from the concentration camp. Fearing sabotage by the concentration camp workers, the Gestapo planted spies throughout the factory. Failures among the rockets being fired in Poland, and the discovery of the air-burst problem heightened fears the rockets were being sabotaged. German workers were not even allowed to converse with the inmates unless there was an SS guard present. However, when the rate of known failures among rockets manufactured at Peenemünde was compared to those of Mittelwerk there was very little difference.

In any event, the use of slave laborers was determined to be very inefficient, and the SS began replacing them with skilled German workers. By the end of the war only 2,000 concentration camp inmates remained on the work force at Mittelwerk.

In the first five months the factory was in operation production climbed, from 50 missiles in January to 437 in May. There was a particularly sharp jump from April (123) to May (437). Then problems began to show up during test firings in Poland, and production slowed while the engineers solved them. Large scale production resumed in August (374 missiles), and continued until the end of the war. From September 1944 until February 1945, Mittelwerk produced more than 600 missiles each month. This was still less than the original production targets, but given the state of German industry, it was remarkable. Finished missiles were shipped via railroad to supply dumps near the firing points.

Originally, Mittelwerk was to be an assembly center for subassemblies manufactured by various contractors. Advances by the Russian Army and the Allied bombing offensive forced these plans to change. Between the occupation of many factories by the Russians, and the bombing by the Americans and British, Germany's arms producing capacity began to deteriorate. To compensate for

View of the cross tunnel, where propulsion sections (left) and parts (right) were stored. *Source: U.S. Strategic Bombing Survey.*

Chapter 5: Dora

this loss of capacity, a plan was implemented to completely eliminate A-4 subcontractors, and fabricate the entire missile at Mittelwerk. In order to accommodate the increased fabrication, second floors were added to some of the cross tunnels.

Elements of the United States Army occupied Mittelwerk in April 1945, and found about 250 rockets in various stages of completion on the assembly line. Large quantities of components were also found stacked up outside the factory. No complete rockets were found. Since Nordhausen was in the portion of Germany scheduled to be placed under Soviet control, there was a hurried effort to evacuate as much materiel from the underground factory as possible. Ultimately 640 tons of equipment, requiring 300 railroad cars, was taken to Antwerp and loaded upon Liberty Ships for shipment to the United States.

The British also explored Mittelwerk prior to its being turned over to the Soviets. Sir Roy Fedden of the Ministry of Aircraft Production headed the technical mission, which arrived in Nordhausen on June 19, 1945. His report of the trip includes descriptions of the beautiful local scenery and lovely weather. These descriptions were included because Mittelwerk "made such a profound impression upon the members of the Mission." Their impressions of the factory are best described by quoting directly from their official report:

"The Mission had been told that Nordhausen (sic.) was a large underground factory, and that they would see extraordinary production methods, but they had no idea that they would be brought face-to-face with such an undertaking. The reaction of the Mission

Tail units being stored outside Mittelwerk. Note that the components were pre-painted in a splinter camouflage pattern. *Source: U.S. Strategic Bombing Survey.*

Space limitations inside Mittelwerk made it necessary to store components outdoors. This photo shows body shells in outdoor storage. *Source: U.S. Strategic Bombing Survey.*

Germany's V-2 Rocket

Entrance to one of the main tunnels at Mittelwerk. Pressure spheres and body sections for the Fi-103 are stacked to the left. Camouflage netting above the tunnel hid it from Allied reconnaissance planes. *Source: U.S. Strategic Bombing Survey.*

to this visit...was one of the utmost revulsion and disgust. This factory is the epitome of megalomaniac production, and robot efficiency and layout. Everything was ruthlessly executed with utter disregard for humanitarian considerations. The record of Nordhausen is a most unenviable one, and we were told that 250 of the slave workers perished every day, due to overwork and malnutrition. Some of the Mission visited the slave workers' encampment, talked to a Dutch doctor who had been there throughout the war, and saw many of the wretched inmates, who were in an appalling state, although receiving every medical attention now. They also saw stretchers heavily saturated in blood, a room in which the bodies were drained of blood, and the incinerators in which the bodies were burnt. These are all facts which require to be seen to be fully appreciated. This terrible and devilish place has now passed into Russian hands, and it is sincerely hoped that our allies will deal with it in a proper and adequate manner." (Fedden, Roy, *The Fedden Mission to Germany.*)

A-4 (V-2) Missile Production

Month	Number	Unit Cost in RM
January 1944	50	100,000
February	86	100,000
March	170	100,000
April	123	100,000
May	437	100,000
June	132	90,000
July	76	90,000
August	374	90,000
September	629	80,000
October	628	80,000
November	662	70,000
December	613	60,000
January 1945	690	60,000
February	617	50,000
March	540	50,000
Total for 15 Months	5,947	

Source: V-2 Rocket Attacks and Defense.

Tunnel entrance at Mittelwerk after the war. Railroad tracks for transporting missiles from the factory lead out of the tunnel.

6

Field Trials and Preparations

The need for training personnel was identified even before the first successful flight, so a Training Command was created within the VKN in April 1942. This command was tasked with testing and evaluating the A-4 and its field handling equipment, developing tables of organization and equipment for future rocket batteries, and training instructors for the units. The training unit was headquartered in Karlshagen, a little south of Peenemünde East.

The initial training course was divided into three phases. Phase A.I lasted eight days, and provided a basic overview of the missile. The second phase comprised two parallel seven week courses, Phases A.IIa and A.IIb. Phase A.IIa covered electronics training; A.IIb addressed other technical aspects of the missile. Oral and written examinations followed the second phase. Based on their examination results, individuals were selected for further training in the shops and laboratories at Peenemünde, or were removed from the program. Thus, the early training comprised lectures, followed by an apprenticeship in the shops and laboratories of HVP.

Training continued in this manner until Hitler raised the A-4 priority rating in July 1943. After meeting with Hitler, Dornberger issued the orders to create a formal training unit, the *Lehr und Versuchs Batterie* 444 (The Test and Experimental Battery 444). This unit comprised military personnel from the VKN, and was charged with developing the tactical doctrine for the A-4, training operational crews, and conducting final firing trials before the missile was committed to combat. *Oberst* (Colonel) Gerhard Stegmaier commended the 444th.

About a month after the creation of the *Lehr und Versuchs Batterie* 444, Dornberger submitted a proposal for the organizational control of the A-4. Within the proposed organization he would command the field organization, and would have a staff similar to other divisional field artillery commanders. He also requested that he be relieved of his duties as head of *Wa. Prüf* 10 and 11, and be appointed *Beaufragter zur besonderen Verwemdimg des Herres*, or Long Range Weapons Special Commissioner for the Army, abbreviated BzbV Heer. On September 1, 1943, General Dornberger was appointed both Senior Artillery Commander 191 and BzbV Heer. His duties in the latter position were to:

1. Accelerate the final deployment of the A-4 and its field equipment
2. Establish the supply system necessary to procure necessary raw materials and equipment
3. Raise and train units
4. Supervise mass production
5. Conduct field trials
6. Make all necessary preparations in France for the commitment of the missile to combat

Dornberger set up his headquarters near Schwedt, on the Oder River. He organized his staff into a command group, a supply group, and an engineer group. Prior to his appointment as Senior Artillery Commander 191, Dornberger had only commanded one military unit, a small solid-fuel rocket battery at Kummersdorf when he was a Captain.

The first operational unit, *Artillerie Abteilung* 836, was formed in October in Vierow, east of Lubmin. The following month a second unit, *Artillerie Abteilung* 485, was activated in Naugard. Both units were motorized, and were approximately equivalent in size to an artillery battalion. Each had three firing batteries, a headquarters battery, and a supply and service battery. There were three firing sections, with one missile launcher apiece within each battery, giving each *Abteilung* nine launchers.

Oberst Stegmaier became commandant of the Long Range Artillery School, in addition to his duties as commanding officer of the 444th. He also commanded the replacement unit for the field *Abteilungs*, the *Artillerie Ersatz Abteilung* 271, or 271st Replacement Artillery Battalion. Stegmaier established his headquarters in Köslin, where he could supervise the various activities.

The quality of instruction at the Long Range Artillery School had improved greatly during the past year. Instead of lectures followed by apprenticeship in the shops at Peenemünde, which had been the initial practice, students at the school trained with elaborate mock ups, cutaway models, and practical exercises with real field equipment. In addition to the "A" course, which was intended for officers and those with engineering backgrounds, there was a

Field preparations of an A-4 missile. The way the missiles were constructed—that is, using a sheet steel skin spot-welded to a frame—created a "dimpled" appearance.

Chapter 6: Field Trials and Preparations

"B" course for those not slated for positions of authority. The "B" course lasted about six weeks, and contained six phases: B.I, propulsion system; B.II, steering mechanisms; B.III, thrust termination control; B.IV, Leistrahl guide beam apparatus; and B.V, fuels.

The new A-4 units trained in Köslin for the classroom portion of their instruction, then moved to Poland, at the proving grounds provided by Himmler, to practice handling and launching missiles. Construction of barracks, vehicle shelters, and other facilities at the Heidelager took place during September and October. A rail spur connected the site to the Cracow-Lemburg line. The Polish Underground noticed this activity, but still did not have any idea of what was being built there, because the launch site itself was enclosed with a double fence of barbed wire. Still, they observed what they could and sent reports back to London.

The 444th launched the first A-4 in Blizna on November 5. The launch did not go well.

It was a frigid 14° when the crew prepared the rocket, and the ground was frozen. Lacking experience in launching the rocket under field conditions, they assumed the ground was hard enough to support the portable launch platform, and it was at first. However, during the preliminary stage, the rocket's exhaust impinged on the ground. This thawed the ground, and one of the legs of the launch platform sank in the mud. When the main stage order was given, the rocket was leaning noticeably. The guidance system could not correct the list once the A-4 lifted off, and it crashed a short distance away.

While firing trials began at Blizna, the *Wehrmacht* Operations Staff under General Alfred Jodl wrestled with how best to deploy both the A-4 and Fi-103. They finally decided to field the weapons under a unified, joint services field command. Hitler signed the necessary directive on December 1, 1943, to create the *LXV Armee Korps zbV*. The letters zbV stood for "zur besonderen Verwendung," or "for special employment." An Army general, 62-year old Erich Heinemann, was selected to command the LXV Corps. Since this was a joint services command, his Chef of Staff, *Oberst* (Colonel) Eugen Walter, came from the *Luftwaffe*. The rest of the staff was divided as evenly as possible between the Army and *Luftwaffe*.

Heinemann was the former Commandant of the artillery school. He had considerable experience developing tactical procedures for new artillery weapons. Field units under the LXV Corps comprised the 155th Flak Regiment for the Fi-103, and the A-4 units under the command of Artillery Commander 191. *Luftwaffe* Colonel Max Wachtel continued to command the 155th Flak Regiment as it prepared to enter combat. Corps headquarters was established in France at St. Germain.

Battery firing trials at the Heidelager, near Blizna, Poland, c. 1943.

Dornberger also moved his headquarters to France, near Maison Lafitte, but he spent most of his time at Peenemünde, or in Poland, dealing with developmental problems. Under his dual assignments as BzbV Heer and Artillery Commander 191, Dornberger controlled both the A-4 development and field deployment, and he believed this was the best way to bring the missile to operational status. General Heinemann disagreed with Dornberger that it was best to have both realms under a single authority. The *Wehrmacht* Operations Staff agreed with Heinemann, who also felt an officer with field experience should command the troop units. On December 29, 1943, Heinemann replaced Dornberger as Artillery Commander 191. *Generalmajor* Richard Metz, a veteran of the Russian Front, succeeded Dornberger in the post.

Dornberger retained the post of BzbV Heer, which left him with plenty to do. Serious problems with the A-4 cropped up during the flights at Blizna. By the end of March flights had been attempted with 57 rockets, none of them with live warheads. Fewer than half (only 26) left the ground at all, and then only four actually impacted within the target area. With more launchings this improved slightly. Even with the improvements, the German rocket scientists found that fewer than 20% of the missiles successfully reached their target.

They observed three types of failures. Sometimes the engines quit less than 100 feet above the launch pad, while others reached altitudes of 3,000-6,000 feet and exploded; still others traveled the full range, only to explode a few thousand feet above the ground. At first sabotage was suspected, for some of the missiles were built by slave labor at the underground factory Mittelwerk.

There was sufficient reason for this suspicion. The first four rockets off the assembly line, serial numbers 17,001-17,004, contained hundreds of defects, and had to be returned to the factory for rework. Even with the second pass through Mittelwerk, the missiles still were not satisfactory. On January 27, 1944, missile number 17,003 was the first one produced at Mittelwerk to be launched. It crashed and exploded two seconds after lift off. As more missiles from Mittelwerk were flown, it was possible to compare their failure rates with those built at Peenemünde. Both groups had nearly identical failure rates, which eliminated the sabotage theory. Something was wrong with the missile design.

An intensive examination of all components turned up the reason for the early engine shut downs. One of the numerous relay switches in the missile proved overly sensitive to vibration from the rocket motor. Once its design was improved this particular problem vanished. The explosions during powered ascent proved a little more difficult to track down, but were eventually traced to the fuel distribution lines and combustion chamber connections. Engine vibrations loosened the connections, and permitted fuel spillage in the tail unit. Redesigning the connections at each end of the pipes, as well as the way they were installed, solved this problem.

However, nearly 70% of the rockets still exploded over the target. Investigating this problem nearly cost von Braun his life. Seeking first-hand observations of the problem, von Braun stationed himself precisely at the predicted impact point, reasoning this was the spot least likely to be hit. One day von Braun stood in an open field waiting for an incoming missile. Looking in the direction the rocket was coming from, he saw a thin contrail heading right for his position. At the last possible second von Braun realized this was going to be one of the few successful rockets. He dove for cover just before the rocket hit nearby. The explosion hurled him into a nearby ditch—the rocket hit just 300 feet from where he had been standing.

At another launch, Dornberger observed an incoming missile through high-powered binoculars. In the blink of an eye he noticed a cloud of steam surround the rocket as it broke apart. Dornberger believed this was residual oxygen released from the tank as the rocket disintegrated. He also thought he saw the rocket yaw almost 20° just before the cloud formed, but suspected this may have been an optical illusion. What he was sure of, though, was that the warhead and control section broke cleanly away from the rest of the rocket.

During reentry, the skin of the rocket reached 575° F. Von Braun and his team found that placing glass wool insulation between the propellant tanks and the body shell raised the success rate to 70%. Further minor modifications to the control section and structure reduced the failure rate by another 10%. Even with this improvement, nearly 1 in 5 rockets still exploded in the air above the target. Finally, by the summer of 1944, the problem was virtually eliminated by adding a second layer of skin over the forward portion of the midsection and strengthening the structure.

While he wrestled with the technical problems of making the A-4 operational, Dornberger also had to wage a political battle with Himmler, a battle with potentially deadly consequences. Himmler's aspirations went beyond simply controlling the manufacture of the A-4 and having the missile tested at an SS camp. He wanted full control over the weapon, and was about to demonstrate his strength within the Nazi hierarchy.

Himmler tried approaching von Braun directly, summoning the young missile scientist to his headquarters, where he offered his personal backing and support if the rocket program were placed under SS control. Von Braun told the *Reichsführer* the A-4 was like a fragile, delicate flower that needed careful nurturing and tending to blossom. Perhaps somewhat imprudently, von Braun said Himmler's offer of backing would be like trying to fertilize that delicate flower with a fire hose of manure. Himmler's tone turned icy, and he dismissed von Braun.

On March 15, 1944, Himmler had von Braun, Klaus Reidel, and Helmut Gröttrup arrested for sabotage. According to Himmler, because they speculated about the future uses of rockets at a party, they obviously were not devoting their full energies and attention to the German war effort. This made them saboteurs. Dornberger unsuccessfully sought their release through normal Army channels. He finally managed to have Speer intervene with Hitler, pointing out that the loss of the three would irrevocably cripple the project. Hitler appeared annoyed, because he had to countermand his beloved *Reichführer* SS, but assured Speer that von Braun would be "protected from all prosecution as long as he is indispensable, difficult though the general consequences arising from the situation" were. Although Dornberger appeared to have won, Himmler's message was clear—nobody in the A-4 project was beyond his reach as he sought control over the effort.

Chapter 6: Field Trials and Preparations

In the Spring of 1944 General Metz personally inspected the A-4 units. He was appalled. The 836th was the best trained and equipped, but its personnel still needed more practical training before they could be deployed. The 485th was under strength and under equipped. A third unit, the Artillery *Abteilung* 962, was not even inspected, because it was due to be disbanded, and its personnel reassigned to the other batteries. While Metz was impressed with Stegmaier, he was very critical of the battalion commanders. One of them he considered useless, idle, talkative, and a fool; the other commander simply wasn't up to the job.

For security purposes, Dornberger issued an edict that soldiers were to be taught their specific tasks and nothing more. Only the engineers from Peenemünde possessed the "big picture" regarding procedures. Metz realized this would prove crippling under operational conditions, and urged Dornberger to rescind this order. General Metz believed that ordinary soldiers could be trained to launch the A-4, and that specialized engineers were only necessary during training, and as advisors during field operations.

Following his inspection, General Metz reported to General Heinemann that, at its current state of development, the A-4 was a menace to the troops launching it, and those beneath its flight path. He felt Dornberger's predictions for the delivery of ground handling equipment for the firing batteries was too optimistic, and that further refinements were needed in the training program. He also believed the organization was too small to warrant a General as commanding officer. Metz was so disheartened that he asked to be reassigned somewhere else. Heinemann turned down his request, and directed him to continue preparing for the A-4's deployment.

A three day exercise was held at the Heidelager May 18-20 to assess the readiness of the A-4. General Metz determined it would take at least two to three months before the missile could be deployed. Planning began for the opening attack with the A-4 against England. It was code named Operation Penguin. The Army High Command issued orders for launch sites in northern France to be ready by the end of June or early July. A major storage location for completed missiles was created in the caves near Mery sur Oise.

Then, the Allies invaded France on June 6, 1944. The immediate effect of the D-Day invasion on the A-4 was the stoppage of all work on missile storage and launch sites in France. However, work continued on the Fi-103 sites, which were almost ready when the Allies landed in Normandy.

7

V-Weapons

Germany unleashed the Fi-103 on June 13, 1944. The Nazi Propaganda Ministry dubbed it the "Vergultungswaffe-1" (Vengeance Weapon-1), or more simply, the V-1. Hitler believed the V-1 could terrorize the British into surrender.

During the Summer of 1943 strange looking structures started showing up along the French coast. By late September, the Germans had 40,000 workers from the Todt Organization working on the sites. The British labeled them "ski sites," for that is what they looked like. Actually, they were launch catapults for the Fi-103. Although the Allies were not sure of their purpose at first, they noted one ominous characteristic; all of the sites pointed directly at London. There were 96 launch sites: 64 main and 32 reserve. Convinced that they were some sort of weapon, the Allies began bombing the "ski sites" on December 5, 1943.

With their large ramps and supporting buildings, the first sites were easy to spot. The German response to the bombing was to create smaller, more easily camouflaged launch ramps. These were harder to detect, particularly since the Germans began using existing farm buildings near the ramps to store and prepare the cruise missiles. Not only were the modified sites harder to detect, they could also be assembled quickly. One *Luftwaffe* construction crew erected a site in 18 hours. The Allied bombing slowed the deployment of the V-1, but could not stop it.

When the V-1 arrived, the British responded to the threat by deploying antiaircraft guns, fighters, and moored balloons around London. At first most of the V-1s, which became known as "Buzz Bombs," got through, but as tactics evolved, fewer than 10% got past the defenders. The distinctive sound of the pulse jet engine was the source of the nickname.

Throughout the summer of 1944, the Germans launched thousands of Buzz Bombs against London. Early in the war Londoners lived through the "Blitz," as the Germans prepared to invade England. By mid-1944 German air raids had greatly subsided, and amounted to little more than isolated raids by a few airplanes at a time. Now, Londoners found themselves again being targeted, this time by a "Robot Blitz."

While the V-1s flew across the English Channel, the Nazi Propaganda Ministry warned of an even more devastating weapon, a V-2, which would turn the tide of war in Germany's favor. While the public may not have known what form the V-2 would take, British authorities had a fairly clear picture of the weapon, thanks to the recovery of rockets from Poland and Sweden.

When the Germans began firing missiles in Blizna, the Polish Underground sent frequent reports to London about developments at the Heidelager. The Poles even managed to penetrate the German security to provide information on what was going on inside the compound. Then, on May 20, 1944, a rocket landed in a swampy area near the Bug River. The rocket was nearly intact. Working quickly, the Poles dragged the missile into the river. A sympathetic farmer then herded his cattle through the water upstream so the mud they stirred up would hide the rocket from the German search crews. After a few hours of searching the Germans gave up, and went back to Blizna. That night, the Poles moved the rocket to a barn near Hotowczyce-Kolonia.

A team from the Research Committee of the underground Home Army dismantled and examined the rocket. During the examination, several preliminary reports were dispatched via courier to London. On July 12 they finished their final report, comprising 4,000-word text, 80 photographs, 12 drawings, a map of the Heidelager, and three appendices. The report and eight pieces of the missile were smuggled out aboard a C-47 Dakota on July 28. Sadly, the receipt of this cargo from Poland failed to attract the attention it deserved, because components from another A-4 had arrived only days before.

On June 13, 1944, a missile fired from Peenemünde strayed off course and landed in Sweden. The Swedish government lodged an official protest with the Germans, then agreed to let the British recover the wreckage. The missile broke apart in the air, so pieces were scattered over a wide area. Eventually, over two tons of debris was collected and shipped to England.

Allied technical experts operating under the code name "Big Ben" began assembling the gigantic jigsaw puzzle. With wires,

Chapter 7: V-Weapons

ropes, and supports holding the torn and twisted pieces, the missile began to coalesce. Working around the clock, the Big Ben Committee unraveled the mysteries of this new weapon. By early September they had an amazingly accurate estimate of the rocket, its size, and capabilities.

Overall length was about 46 feet, and the weight appeared to be about 14 tons. The odor of the fuel tank led to the conclusion that alcohol fueled the missile. They deduced that liquid oxygen was the oxidizer, and the propellants were fed to the combustion chamber by a turbopump. Using data from Dr. Goddard's research they calculated the performance of the missile. The range was estimated at between 175 and 200 miles. The group made only one error in their assessment of the missile. This particular round had been used to test the radio guidance system for the *Wasserfall* antiaircraft missile. When analysts found this equipment in the wreckage, it led them to the erroneous conclusion that the A-4 was radio controlled.

Throughout the summer, the Allies discovered further evidence the Germans were on the brink of committing the large missile to combat. Fielding such a weapon required considerable logistical support. To support the field units, the Germans planned to ship missiles from the factory to field dumps throughout northern France. Many of these sites were nearly complete when the Allies invaded on June 6, 1944. As the Allies advanced through France, they found these storage areas.

A main supply dump was found at Hautmesnil, in a quarry next to the Caen-Falaise road. It comprised a series of underground galleries. A railroad spur had been extended from the main track to the site. Other storage sites were discovered in the Normandy area at La Meauffe, Bois de Bougé, and Sottevast. Altogether, these sites provided storage for up to 150 missiles in the Normandy area. Allied troops also found sites in the Pas de Calais area. An aboveground dump was found near Monchy-Cayeux, complete with 40 "rocket trolleys," a gantry with a five-ton winch, personnel quar-

On June 13, 1944, an A-4 missile strayed off course and landed in Sweden. Although officially neutral, the Swedish government allowed the British to recover the wreckage. The pieces they gathered resembled these, which were found in Belgium in late 1944. From the nearly two tons of debris the British collected, they developed an accurate picture of the missile and its capabilities. *Source: Official British photograph.*

Germany's V-2 Rocket

On September 6, 1944, Germany unleashed the A-4, which the Nazi Propaganda Ministry dubbed the V-2, for *Vergultungswaffe Zwei*, or Vengeance Weapon Two. Although the initial attack was unsuccessful, thousands of missiles reached their targets in the final six months of the war in Europe.

Chapter 7: V-Weapons

ters, and wooden shelters for the missiles. This site had a capacity of storing 28 missiles.

The Allies also discovered prepared firing sites throughout the area. Consisting of concrete pads covered with dirt and grass for camouflage, these were a legacy of the first launch at the Heidelager. Seeing how the ground thawed beneath the launch table, General Heinemann was convinced that the missiles could only be launched from prepared sites. Subsequent tests at the Heidelager showed such preparations were unnecessary. Logs or metal plates could successfully support the launch stand, but Heinemann remained adamant in his belief that the A-4 could not be fired from anything but a prepared platform.

Up to the summer of 1944, the Germans clung to the plan of eventually launching missiles from the large bunker near Wizernes. Just as they had done to the bunker at Watten, American and British bombers intervened. Dropping six-ton "Tallboy" bombs on the Wizernes site, they churned up the ground around the dome so badly that further construction was impossible. On 6 July another raid on the surviving portion of the Watten bunker caused its collapse. Despite his penchant for bunkers and grandiose construction projects, Hitler finally decreed on 18 July that the large launching sites no longer had to be pursued. The question of launching by mobile field batteries was finally settled.

July also saw the abandonment of the Heidelager due to Soviet advances through Poland. The A-4 launch site was moved 10 miles from Tuchel. This area was named *Heidekraut* (Heather). Developmental launches and crew training continued from this site.

The July 20, 1944, attempt by a group of high-ranking Army officers to assassinate Hitler had far reaching effects on the missile program. One of the conspirators smuggled a bomb into Hitler's headquarters near Rastenburg, which detonated during a briefing. Hitler survived, but was badly shaken. Immediately following the blast Himmler organized his SS to block the coup and keep Hitler in power. He also rounded up the conspirators. In the wake of the attempt, which the Army officers had named "Operation Valkyrie," Himmler emerged stronger than before because of his loyalty to Hitler. Even Albert Speer, who had been one of Hitler's most trusted lieutenants, and was a potential successor as *Führer*, fell under suspicion, because his name appeared on a proposed post-coup organizational list. Hitler rewarded those he saw as loyalists, particularly Heinrich Himmler. The *Reichsführer SS* became the Commander of the Replacement Training Army. From this position he finally seized control of the A-4 program.

On the first day of August, Peenemünde became the *Electromechanische Werke, Karlshagen* (Electromechanical Industries, Karlshagen), and Paul Storch, a former director of the Siemens Company, was appointed its general manager. Dornberger was transferred out of the Peenemünde organization. The German Government retained ownership of the facilities at Peenemünde; the *Electromechanische Werke* operated the development plant under contract.

A week later, Himmler appointed Hans Kammler SS Commissioner General for the A-4 Program. The missile program was taken away from the LXV Corps and placed under Kammler's direction. Already involved in the creation of Mittelwerk and Dora, Kammler had become a frequent visitor to the Heidelager, so he was familiar with the field operations of the missile. Kammler hastily set up a headquarters to oversee the A-4. Although Kammler reportedly despised Dornberger (and vice versa), the latter was the senior most person he could turn to for help, so he kept him on his staff.

On 29 August Hitler ordered Kammler to begin Operation Penguin immediately. Kammler set up his improvised headquarters in Brussels, and issued orders for the firing batteries to be ready by 5 September. Within a few days, 5,306 personnel and 1,592 vehicles were enroute to establish firing positions between Antwerp and Malines. The A-4 was about to become the V-2.

8

V-2 From the Inside

The first V-2 fired in combat was a very different missile than the one that flew on October 3, 1942. In the ensuing 23 months more than 60,000 engineering changes had been made. The V-2 comprised five major subassemblies: warhead, instrument section, mid body, tail unit, and propulsion unit.

The warhead, which was filled with 1,650 pounds of the high explosive Amatol, looked simple, but required considerable effort to develop. Unlike conventional bombs, it struck the ground at supersonic speeds. This complicated the fusing system, because it had to detonate the warhead before it buried itself, yet be insensitive to the rigors and vibrations of launch and reentry. The explosive filler also had to meet similar constraints. Amatol, a mixture of 60% TNT and 40% ammonium nitrate, offered the best combination of explosive power and insensitivity to shock and heat. The 6-millimeter thick steel casing weighed 550 pounds, which left 1,650 pounds for the explosive filler in its 16.7 cubic foot volume.

Two fuses were used: one at the tip of the nose, and one at the warhead base. The nose fuse, designated type KZ-3, had two inertia switches that were perpendicular to one another, and a crushable dome switch. On impact, the inner and outer shells of the dome switch connected and closed the firing circuit. A silica cap over the fuse protected the dome switch prior to impact. The inertia switches would detonate the warhead if it merely grazed the target. As a back up to the nose fuse, a second fuse, designated BZ-3, was at the base of the warhead. It did not, of course, have the crushable switch, but otherwise it was identical to the KZ-3. Closing any one of the five switches in the two fuses initiated the explosive train, and detonated the Amatol.

A pair of switches armed the fuses. The on board timer controlled the first arming switch, and closed it 40 seconds after launch. Thrust termination closed the second arming switch. They had to function in that order. If the engine malfunctioned and shut down

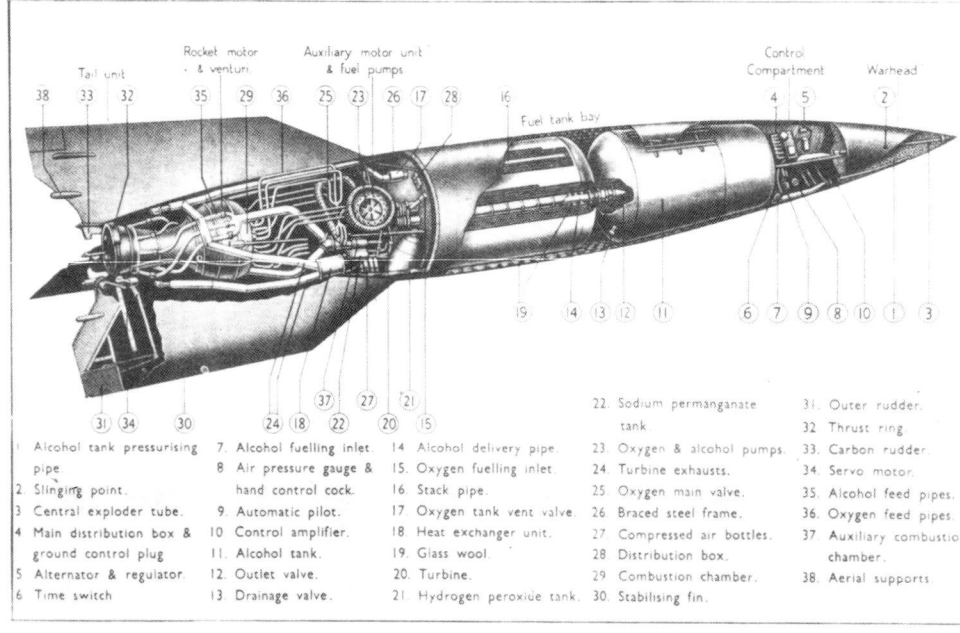

V-2 schematic. *Source: Summary Report of V-2 Control and Stability.*

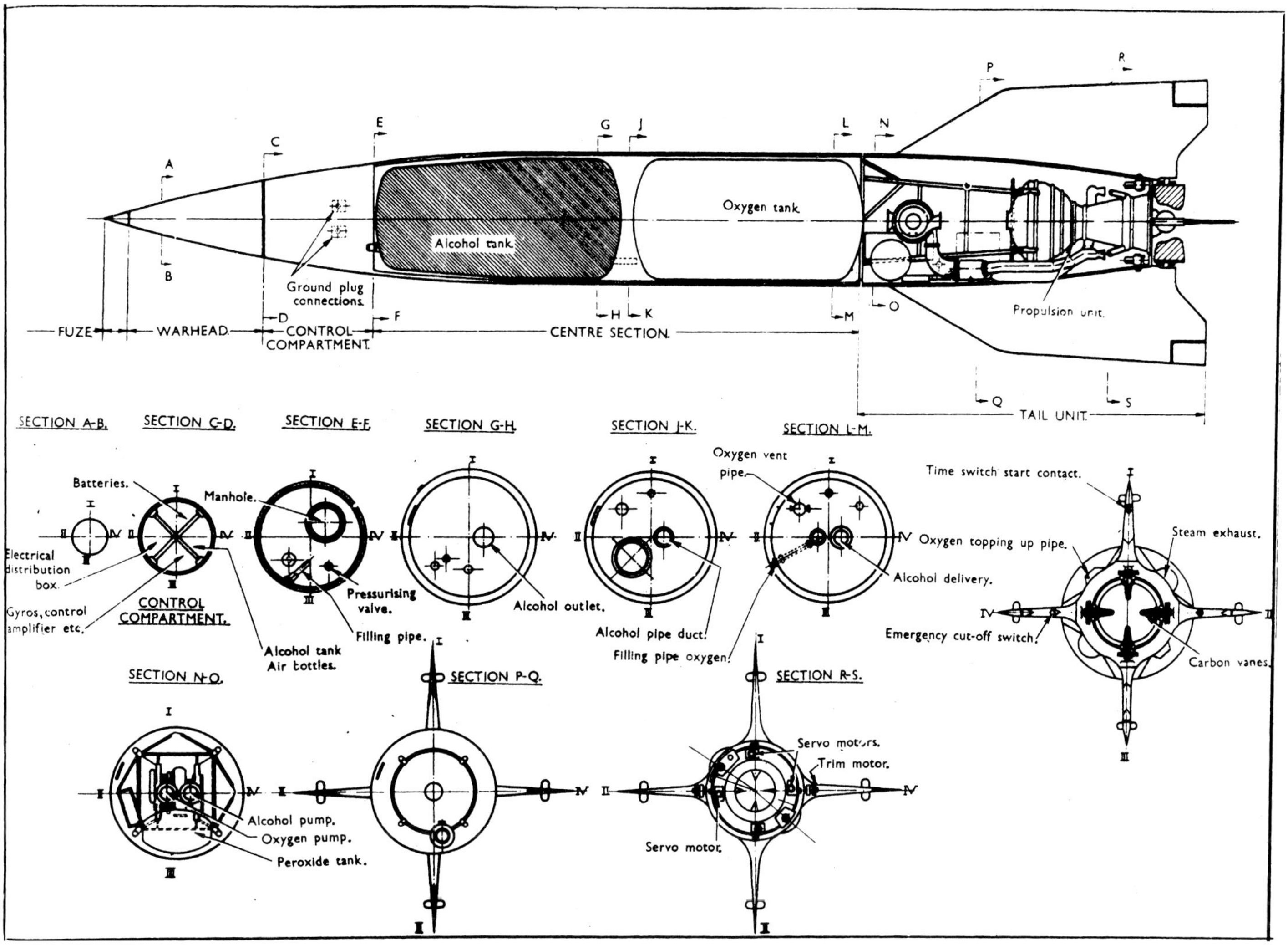

Chapter 8: V-2, From the Inside

V-2 cutaway showing the five major subassemblies: Warhead; Control Compartment; Midsection (identified as "Centre Section"); Tail Unit; and Propulsion Unit. *Source: Summary Report of V-2 Control and Stability.*

Warhead fuses. *Source: U.S. Army photograph.*

THE WARHEAD

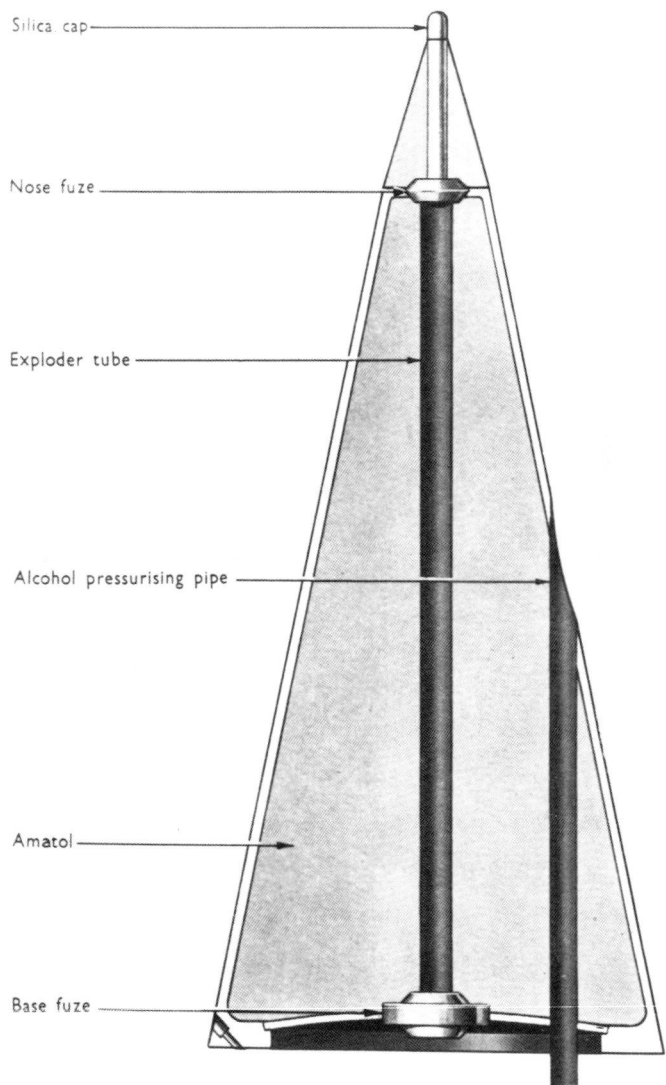

Warhead schematic. Source: Report on Operation Backfire.

before 40 seconds then the fuses did not activate, which prevented normal warhead detonation. This was intended as a safety feature for the launch crews in case a rocket crashed shortly after lift off. When Allied forces overran V-2 launch sites in Holland and Germany, craters found in the vicinity showed the warheads usually detonated anyway. Reliability of the fusing system was excellent—only two unexploded warheads were found during the War.

Each fuse contained a small primer charge, designated F 36. A tube that ran between the fuses, along the length of the warhead, contained the explosive Penthrite. The Penthrite was the detonator. Upon impact, the fuses ignited their primer charges, which detonated the Penthrite, which in turn made the Amatol explode.

There was an opening about halfway down the length of the warhead casing. This opening was for a pipe that extended through the control compartment, and into the alcohol tank. Air entering the opening pressurized the tank, preventing it from collapsing as it emptied, and helping the fuel pump.

The control compartment, a truncated ogive immediately below the warhead, contained the mechanisms that guided the missile during its powered flight. The trajectory of the missile comprised two phases—powered and ballistic. All rockets followed the same trajectory during powered flight. Four seconds after lift off the rocket began a pre-programmed pitch towards the target, until it reached an angle of 47° about 43 seconds into the flight. After this, the elevation angle was held constant until the missile reached the velocity necessary to achieve the desired range. When the V-2 reached the desired velocity the engine was cut off, and the rocket coasted along a ballistic trajectory like any artillery shell.

Two methods of controlling the engine cut off were used. The first was to terminate the thrust by ground command via radio. The other method, and the one used by most rockets early in the campaign, was an on-board device that determined the velocity and issued the cutoff signal.

Radio controlled cutoff proved to be the most accurate, but presented several disadvantages. There was always a fear that the Allies would jam the cutoff signal. It was also feared that the additional vehicles required for the radio equipment would be vulnerable to Allied air attack. Finally, there was the additional logistical

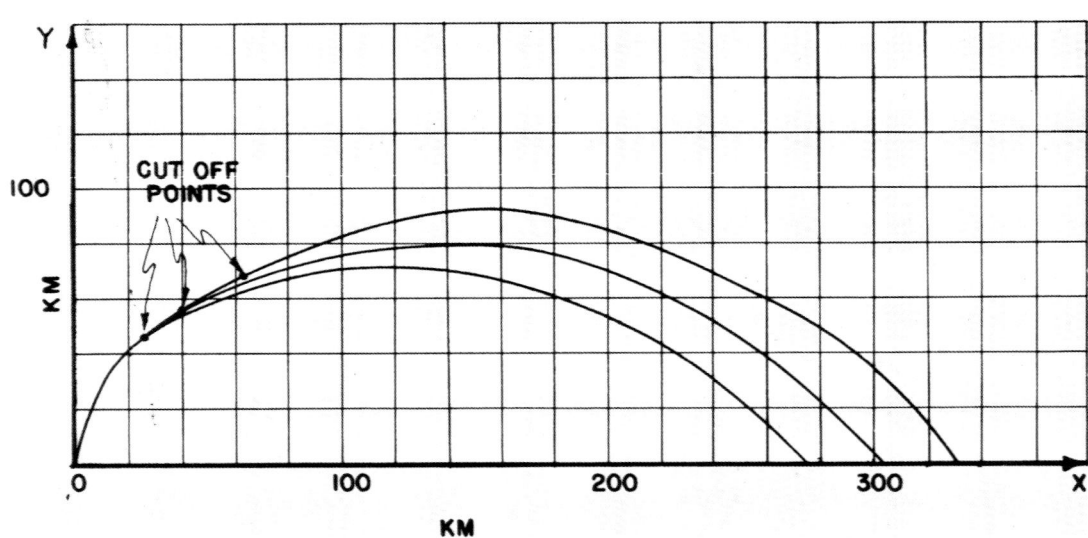

The powered portion of all A-4 missiles was presumed to be the same. Therefore, range became a function of cut-off velocity. Source: Summary Report of A-4 Control and Stability.

Chapter 8: V-2, From the Inside

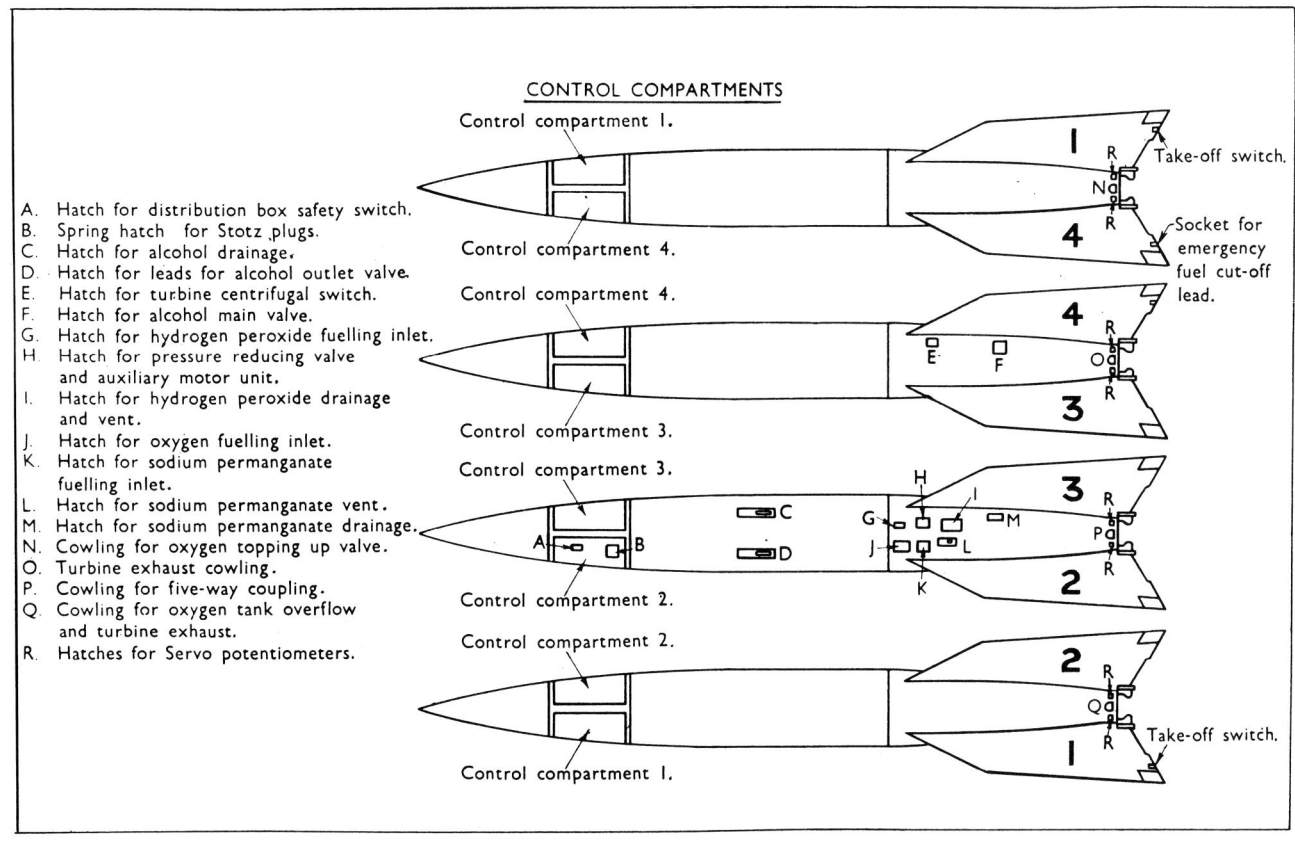

External hatches on the A-4 (V-2) missile. *Source: Report on Operation Backfire.*

burden of maintaining and supplying additional vehicles and equipment.

On-board equipment for radio control consisted of a receiver-transmitter called Verdoppler or Ortley, and a control receiver called Honnef. The Verdoppler received signals on a frequency of 30.7 megacycles, and retransmitted them on 61.4 megacycles. The retransmitted signal was compared to the second harmonic of the one being broadcast from the ground. A beat frequency proportional to the rocket's velocity was produced due to the Doppler Effect. The Honnef comprised a superhetrodyne receiver, an audio-frequency filter unit, a relay unit, and a three-phase power supply. This receiver was used to reduce the fuel supply to the propellant turbopump, and then stop it altogether.

If the cutoff signal was sent to a rocket firing at the full 25-ton thrust level, it took an average 2.7 seconds for the thrust to reach zero, which affected accuracy of the round. Because of this, a two-stage engine cut-off was used. When the rocket reached 95% of its desired velocity, engine thrust was reduced to 8 tons, then terminated completely when the final velocity was attained. This two-step engine cut-off reduced dispersion errors by as much as 50%.

The Germans feared the Allies might learn how to jam the cut-off signals, so they equipped the missile with an anti-jam device called *Kommandogerät*. It used a combination of modulated tones and a receiver with a narrow-band filter. In the event they did de-

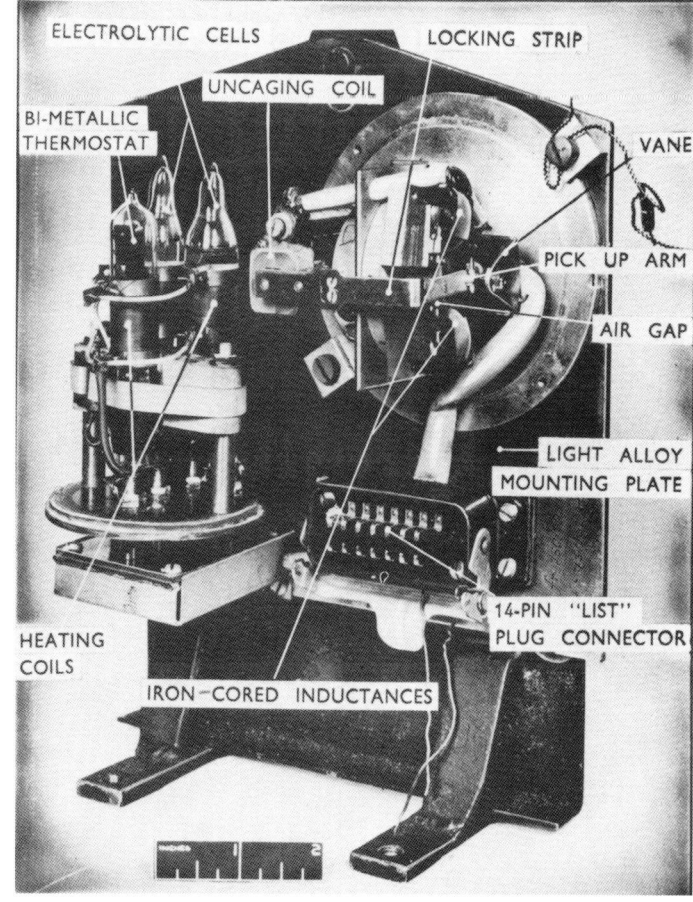

The electrolytic integrating accelerometer, one of the devices used to terminate engine thrust when the desired velocity had been reached. *Source: Summary Report of A-4 Control and Stability.*

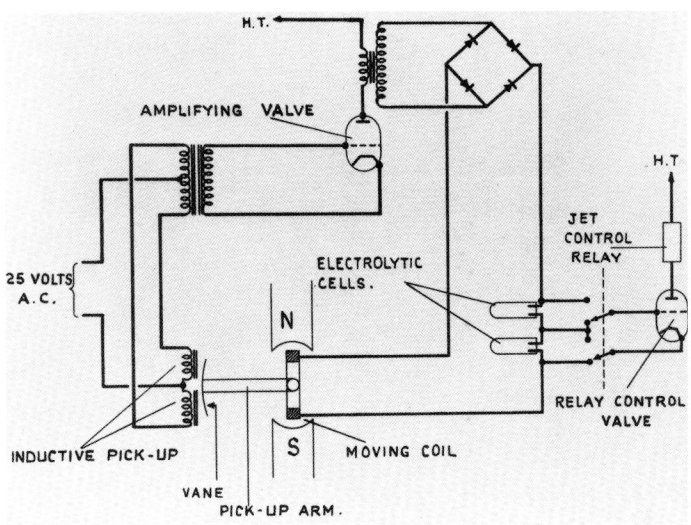

Schematic of the electrolytic integrating accelerometer. Source: *Summary Report of A-4 Control and Stability.*

of silver chloride on the anode) controlled the thrust reduction, while the signal from the first one cut off the engine once the desired velocity was reached.

During powered flight, the rocket continually accelerated, reaching a maximum of approximately 6 g's (1 g is the normal acceleration of gravity, 32.2 feet per second) at burnout. The torque balance accelerometer contained a mass connected to a lever arm that developed a torque proportional to the acceleration. The pivot of the lever arm contained a galvanometer coil that moved through the field of a permanent magnet. Current was applied through the

tect any signs of jamming, the Germans had even more elaborate devices ready. There were no incidents of Allied jamming of the cut-off signal, so these devices were never used.

Integrating accelerometers were the other means of controlling thrust cutoff. These devices determined the rocket's velocity as a function of acceleration and time. When the rocket approached the desired velocity, thrust was terminated in the same two-step manner as with the radio-control cutoff. There were two types of integrating accelerometers: a torque balance accelerometer with an electrolytic integrator, and a gyroscopic type with an accelerometer and integrator.

Professors Theodor Buchold and Carl Wagner of the *Technische Hochschule*, Darmstadt, developed the torque balance accelerometer. It consisted of a device to deliver a direct current proportional to the acceleration, and an electrolytic cell that integrated the current with respect to time.

The electrolytic cell contained two silver electrodes, and an electrolyte solution containing sodium chloride, acetic acid, and sodium acetate. Each liter of solution contained 15 grams of sodium chloride, 60 grams of acetic acid, and 136 grams of sodium acetate. During manufacture, one of the electrodes in the cell was heavily coated with silver chloride. To prepare the cell for use and program it for a desired cutoff velocity, the normal polarity of the cell was reversed so the coated electrode was negative.

With the flow of current reversed, the torque balance produced a current proportional to the Earth's gravitational acceleration, g_o. Current was allowed to pass through the cell for a time period, t_o, such that the product of g_o and t_o equaled the desired cutoff velocity ($g_o \times t_o = V_c$). As current passed through the cell, chloride ions moved to the previously uncoated anode and left a deposit of silver chloride. At the end of the time period the original connections were restored, and the cell was ready for use.

Each accelerometer contained two electrolytic cells. Preflight programming was started in one cell, then the other one about 20 seconds later. In flight, the second cell (which had a smaller amount

The gyroscopic integrating accelerometer. Although not as accurate as the accelerometer with electrolytic cells, this type was more widely used. Source: *Summary Report of A-4 Control and Stability.*

Chapter 8: V-2, From the Inside

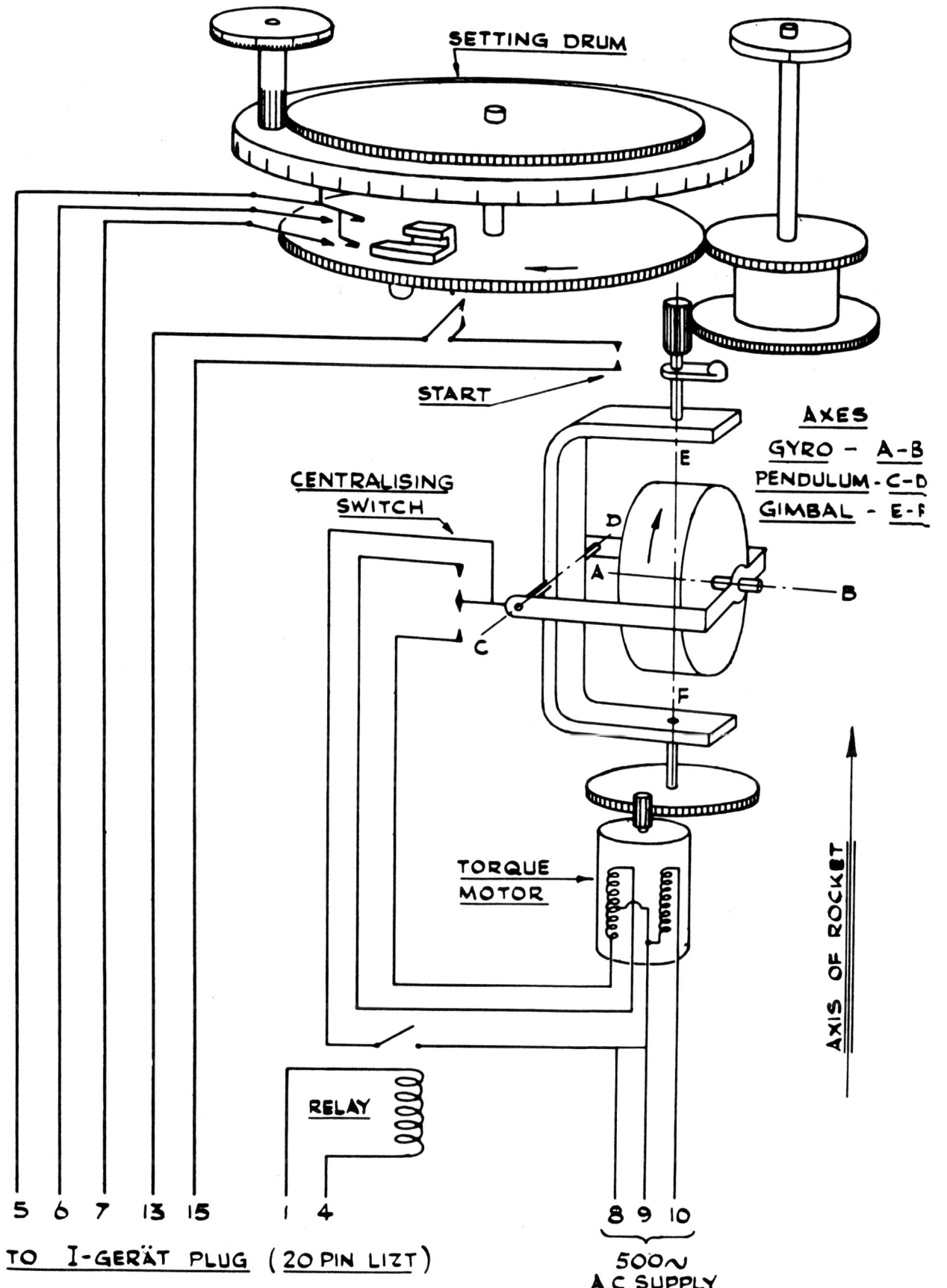

Schematic of the gyroscopic integrating accelerometer. Source: *Summary Report of A-4 Control and Stability.*

coil to counter the torque produced by the mass during acceleration. This current passed through the electrolytic cells.

As current flowed through the electrolytic cells, the freshly deposited silver chloride was transferred to the original electrode. Once the transfer was complete a sharp rise in cell voltage occurred, which triggered the thrust control relays. It was claimed the electrolytic cells controlled thrust cutoff velocities to within 0.1 percent of the desired values.

The gyroscopic integrating accelerometer contained an electrically driven gyroscope suspended from a gimbal that was perpendicular to the axis of rotation of its rotor. The gimbal was balanced, and pivoted about an axis parallel to the long axis of the rocket. Acceleration along the rocket's longitudinal axis caused the gyroscope to precess about the gimbal axis. The angle of precession was proportional to the integral with respect to time of the acceleration, or more simply, the velocity.

Through a series of reduction gears, the gimbal drove a wheel that had three contacts on it. The first contact indicated the start point at lift off; the second controlled thrust reduction; and the third signaled thrust termination. During prelaunch preparations, the second and third contacts could be set at any point on the wheel. Under laboratory conditions, the gyroscopic integrating accelerometer had an accuracy of 0.1 percent; in the field, the accuracy proved to be about 0.3-0.4 percent. Minor variations in the gyroscopic rotors were the main cause of the variations.

Plywood sheets divided the control compartment into four quadrants. Radio control equipment occupied the first quadrant. Quadrant I also contained the batteries that powered the missile's electrical systems. Two 27.5-volt lead-acid batteries with eight cells each connected in series were the main power source. These batteries provided 27 volts with a 20-ampere discharge for about 5 minutes. Both batteries could withstand a 6-g acceleration, and could function in any position.

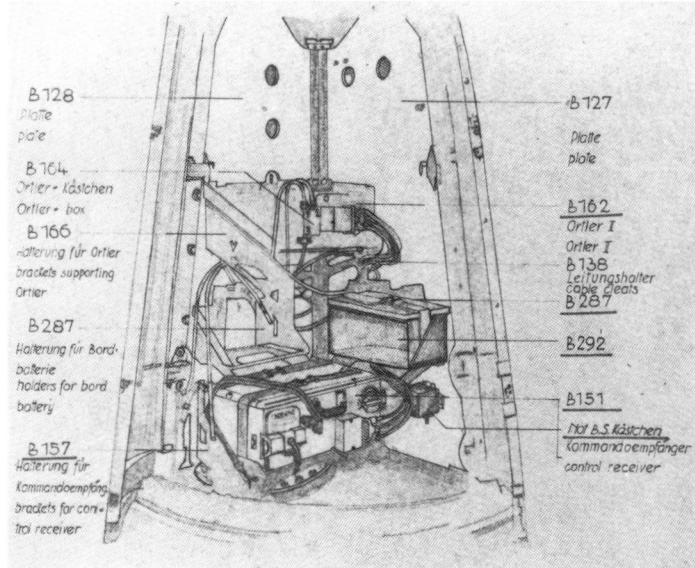

Control Compartment Quadrant I. This compartment held the batteries that powered the missile electrical systems. *Source: Die Fernrakete*

Quadrant II housed the sequence switch, main distribution panel, and fuse arming unit. A mechanical switch controlled functions that had to occur at a specific time and in a specific order. It comprised an electric-motor driven cam shaft that operated a number of contacts. Contacts in the switch controlled the radio equipment, horizontal gyroscope, alcohol tank pressurizing valve, and the warhead arming unit. As a backup to the other fuel cutoff systems, a contact in the switch signaled cutoff 65 seconds after launch. A switch built into fin I started the timer when the rocket lifted off.

The main distribution panel provided a single location for concentrating and cross-connecting leads from the steering controls in the control compartment to the steering mechanisms in the tail unit.

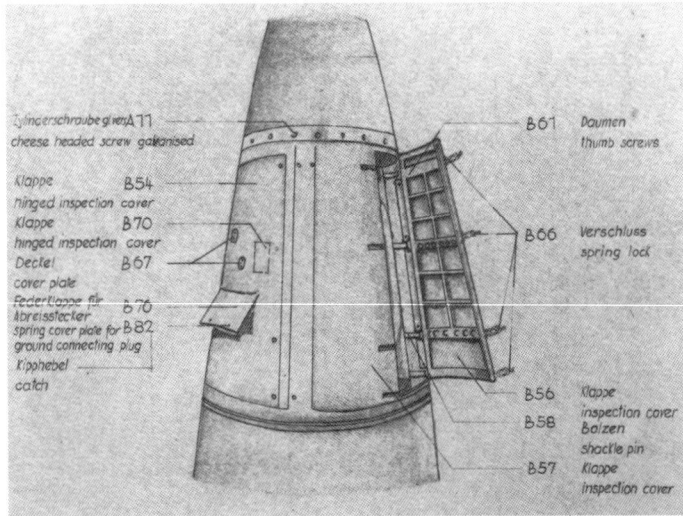

External view of Control Compartment showing latches and fasteners. This drawing is from a manual prepared for the British by captured German personnel after the war. *Source: Die Fernrakete*

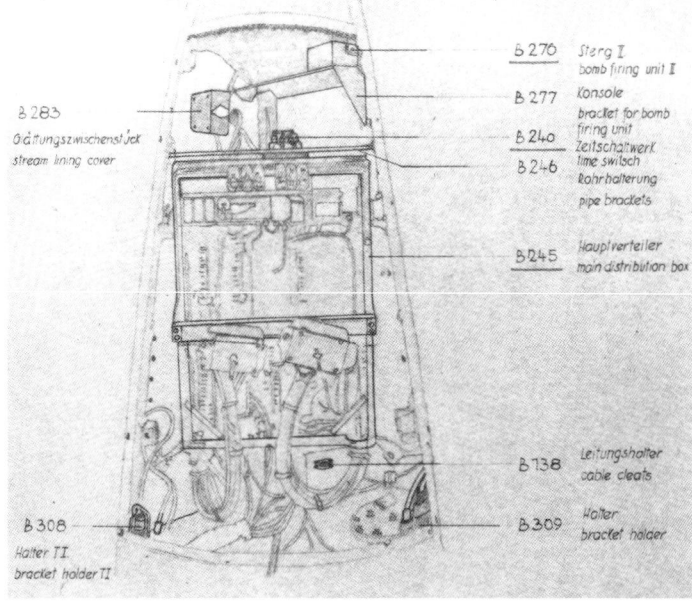

Control Compartment Quadrant II contained the warhead arming unit and main distribution panel for the electrical system. *Source: Die Fernrakete*

Chapter 8: V-2, From the Inside

Power distribution to the other missile systems was controlled through the main panel, as well. Two ground-connecting plugs on the panel provided the interface between the rocket's electrical circuits and ground test equipment.

The third quadrant contained the integrating accelerometer, pitch gyroscope, roll and yaw gyroscope, control amplifier, voltage regulator and alternator, 50-volt signal battery, and the alcohol tank pressurizing pipe and valve assembly. The pitch gyroscope directed the missile during the boost phase. The pitch gyroscope governed the rocket's angling towards the target. A single gyroscope sensed motions in yaw and roll. If any unwanted motions were detected, then corrective signals were sent to the control surfaces in the tail unit.

Both gyroscopes were essentially the same, with rotors being driven by a 500-cycle 3-phase electric motor. Rotor speed was 30,000 revolutions per minute. The difference between the gyroscopes was the addition of a small motor, called the program motor, to the pitch gyroscope. This motor consisted of a ratchet-driven wheel that rotated a small cam shaft through a worm drive. During flight, the program motor rotated the housing of a pick-up potentiometer. Signals from the potentiometer were sent to the pitch rudders in the tail unit, which caused the V-2 to tilt towards its target.

An azimuth-control device called *Viktoria Leistrahl* was contained in Quadrant IV, along with three pressure bottles for the al-

The SG-66 stabilized platform. *Source: Summary Report of A-4 Control and Stability.*

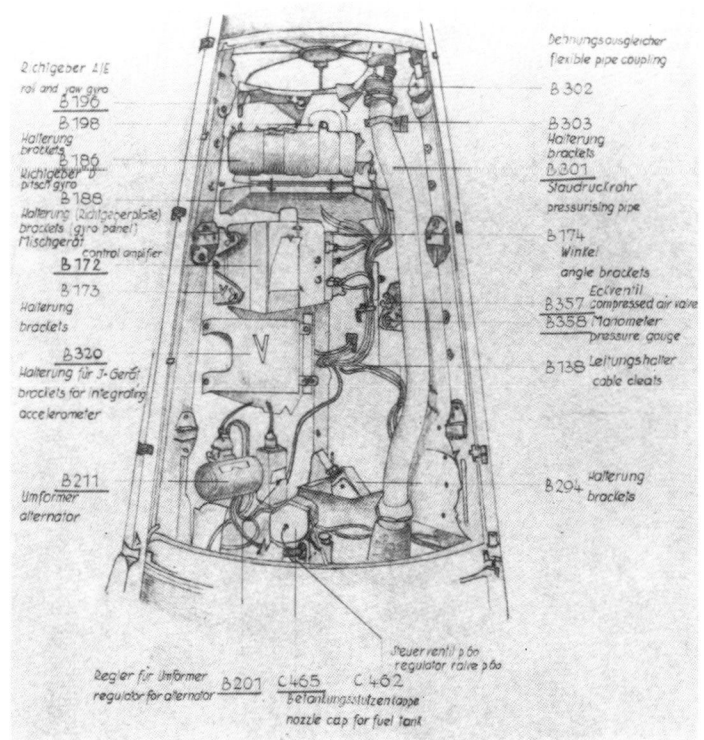

The integrating accelerometers that controlled fuel cut-off and control gyroscopes were mounted in Control Compartment Quadrant III. *Source: Die Fernrakete*

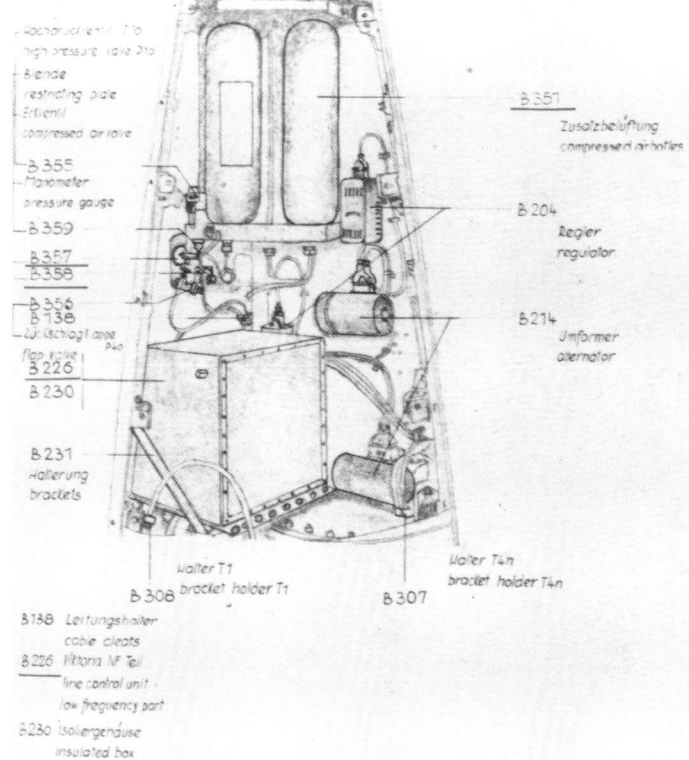

To prevent its collapse as fuel drained, the alcohol tank was pressurized. During the early portion of the flight, this pressure came from a ram-air inlet in the side of the warhead. Forty seconds after launch the air was too thin to pressurize the tank, so the necessary pressure came from three tanks in Control Compartment IV. This quadrant also held the "beam riding" device used mostly on later versions of the V-2. *Source: Die Fernrakete*

cohol tank. The *Viktoria Leistrahl* was a "beam-riding" device. A radio beam was continuously broadcast to the rocket from a ground station. If any deviation from the center of the beam was detected, a corrective signal was sent to the steering mechanism in the tail unit. Developed by the Lorenz Company, the *Viktoria Leistrahl* unit was used mostly on later versions of the V-2.

Although the integrating accelerometers shut the fuel off once the desired velocity was reached, small variations in either engine thrust or the pitch program frequently resulted in sizable errors in the impact point. To reduce the effects of such errors, the Germans developed a stabilized platform that contained three gyroscopes that sensed acceleration and movement about the pitch and yaw axes. Called SG-66, this platform was the forerunner of modern guidance systems. Developed late in the war, only 15 or 20 experimental missiles ever carried the SG-66. When combined with a double integrating accelerometer to more precisely control fuel cutoff, the Germans claimed the SG-66 gave the V-2 an accuracy of 500 yards over a 200-mile range.

Forty seconds after launch the valve in the alcohol tank pressurizing tube closed, and pressure to the tank came from three compressed air bottles in Quadrant IV. Each cylinder had a capacity of seven liters and a working pressure of 200 atmospheres.

The missile midsection consisted of the alcohol tank, oxygen tank, and body shell. Both tanks were made from welded aluminum. The body shell was fabricated in two halves that were bolted together. When assembled, 110-mm wide strips covered the seams between the halves. The shell was a semi-monocoque structure, with a sheet steel skin attached to a framework of strengthening ribs, longerons, and stringers. For additional strength, a second layer of skin was riveted to the forward portion of the midsection. There

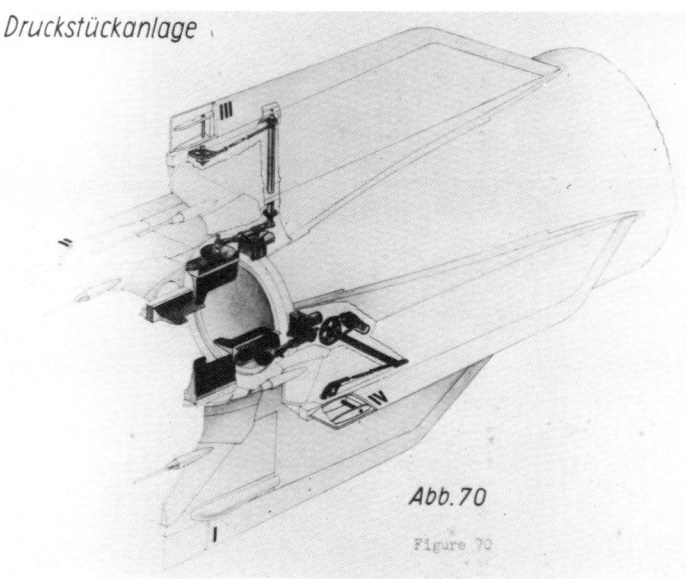

The missile tail unit, showing the system of jet vanes and rudders. *Source: Das Gerät A-4.*

was a layer of glass wool insulation between the tanks and external shell.

The missile tail unit comprised an ogival boat tail with four fins. Control surfaces (jet vanes and air rudders), along with their servo and electric motors, were also contained in the tail unit. Overall construction of the boat tail, which provided an aerodynamic fairing over the propulsion unit, was similar to that of the midsection. A skin of sheet steel was spot welded to a framework of longerons and circular formers. Internal ribs for the fins attached to the form-

A mechanical timer like this controlled the sequencing of missile systems. *Smithsonian Institution photograph.*

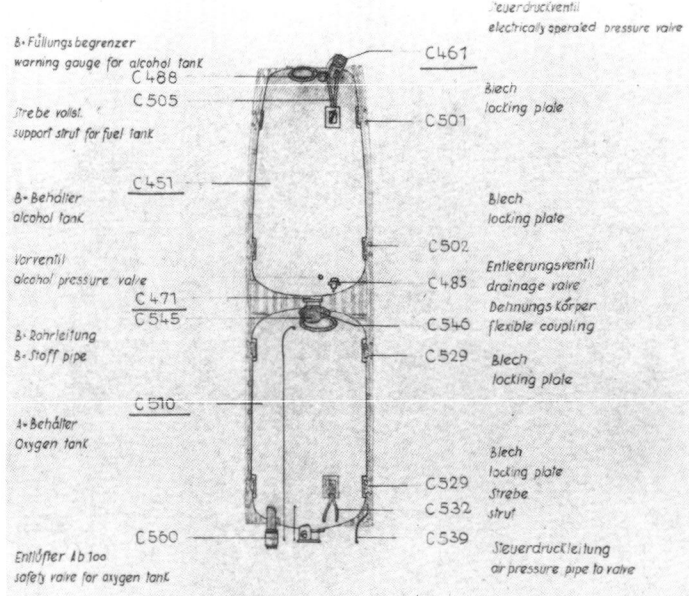

Missile midsection schematic and components. *Source: Die Fernrakete.*

Chapter 8: V-2, From the Inside

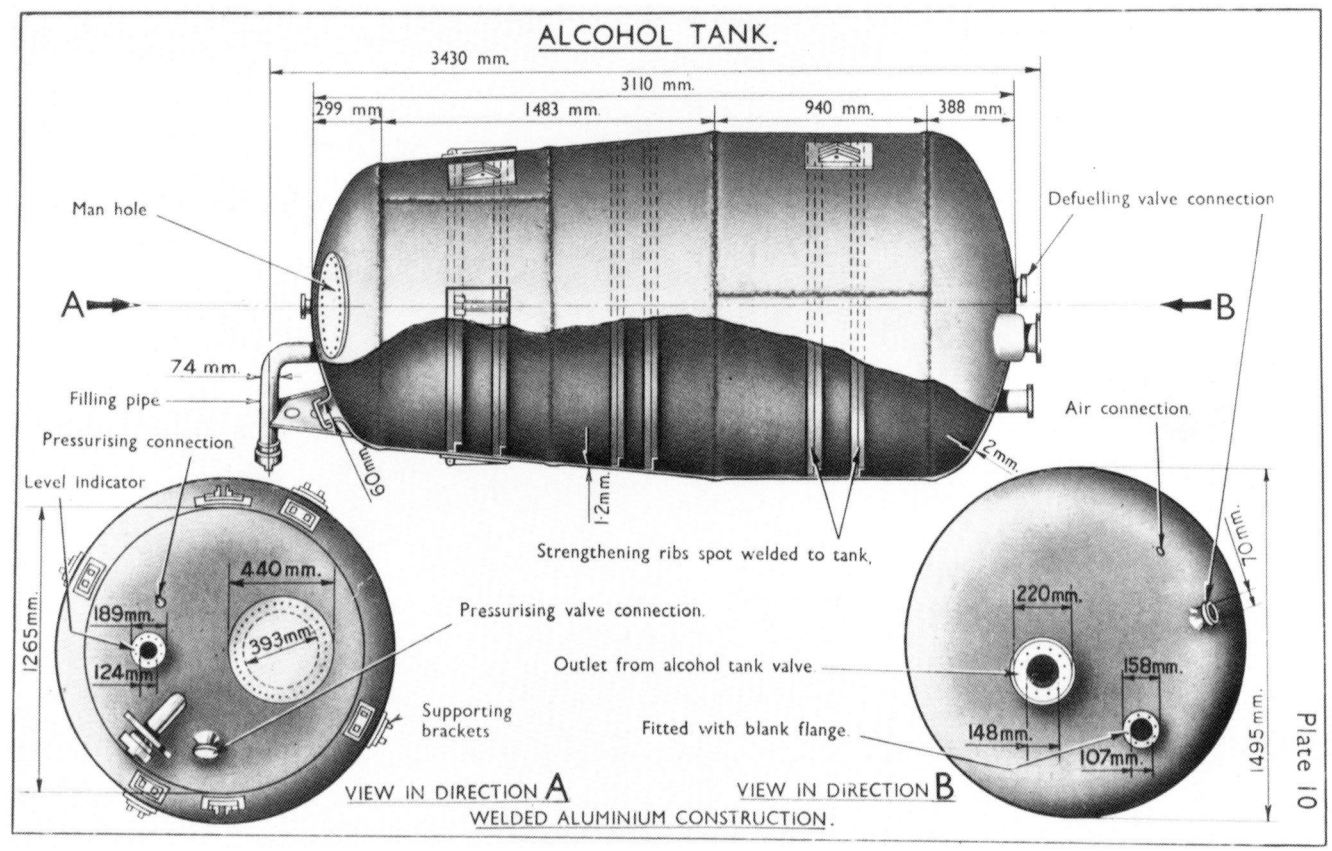

Above: Alcohol tank. Below: Liquid oxygen tank. *Source: Report on Operation Backfire.*

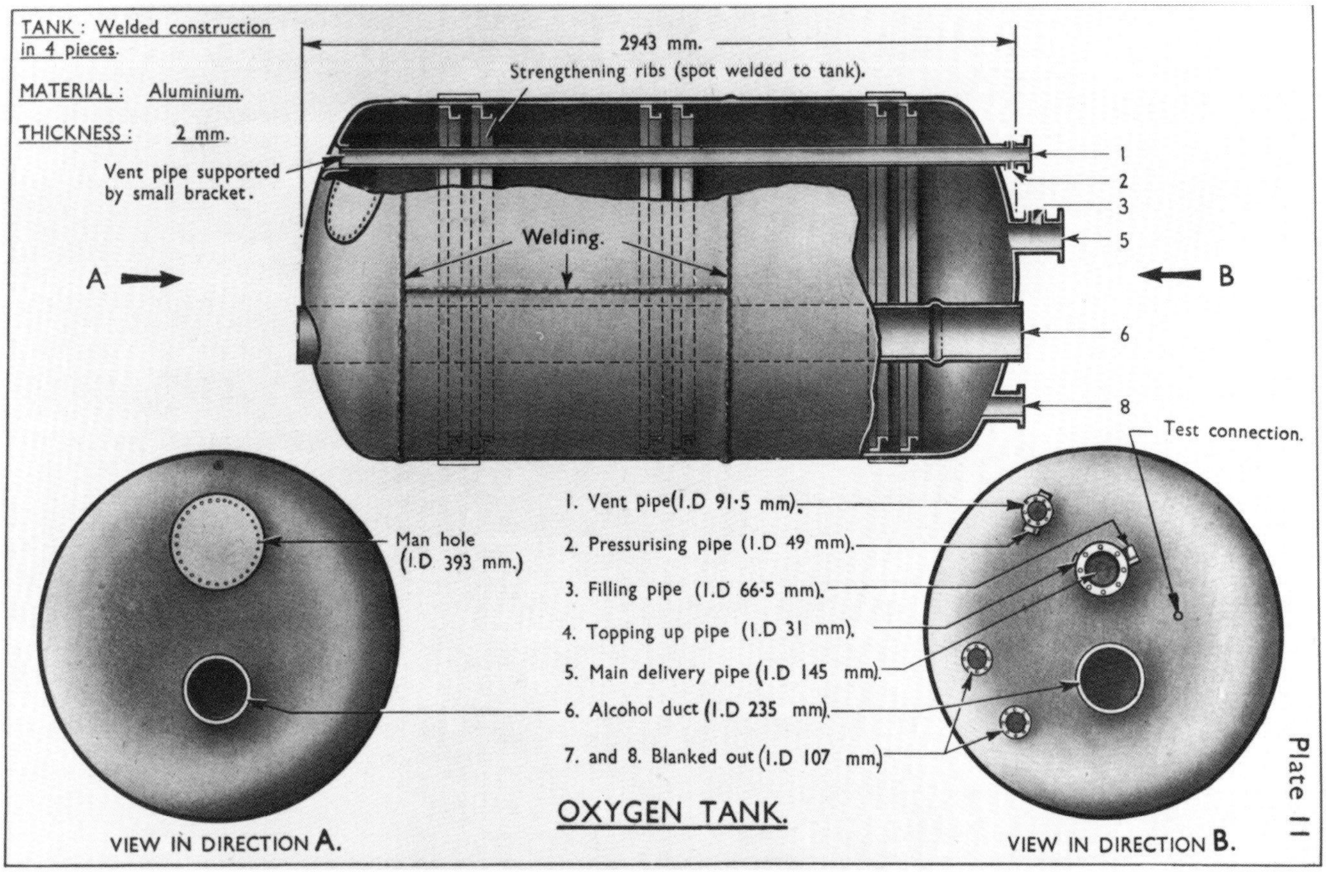

ers. When the midsection and tail unit were bolted together, there was a 5-mm gap between them to allow ventilation through the missile tail. Like the rest of the airframe, the fins were made from sheet steel attached to a framework of longerons and ribs.

Jet vanes in the engine exhaust, as well as aerodynamic rudders on the fin tips, controlled the rocket during powered flight. Movement of the jet vanes deflected the engine's exhaust to steer the V-2. The vanes on fins I and III controlled the rocket's pitch, while the other pair controlled yaw and roll.

The rudders on fins II and IV were linked to the jet vane servo motors by a sprocket and chain, and also helped control yaw and roll. The rudders on fins I and III were not linked to the pitch vanes. Instead, they were driven by separate trim motors, and were brought into operation when the yaw and roll rudders were out of synchronization to correct any unwanted rolling motions.

The jet vanes were made from graphite, which was the material that offered the best combination of heat resistance, cost, ease of manufacture, and availability. Siemens Planiawerke manufactured the jet vanes. Blanks for the vanes were cut in the firm's Berlin-Lichtenberg plant, then shipped to Meitingen, near Augsburg, for machining and finishing.

The blanks were made from the finest available petroleum and pitch cokes. After being mixed with a binder, the mixture was stamped into a horizontal extrusion press. Following extrusion, the blanks were baked at 1,150°C in a ring furnace. The blanks were next X-rayed to check for internal flaws. About 50% of the blanks were rejected at this stage. The blanks were then pitch-treated, baked a second time, pitch-treated again, and then baked for a third time.

The blanks were then shipped to Meitingen for coating with graphite, using a process similar to that used for high-quality graphite electrodes. They were then sawed to rough shape and finished on a "milling-copying" machine that could produce two vanes at a time. It was specially developed for this task, and used tungsten carbide milling cutters. After milling, the vanes were placed in a drying

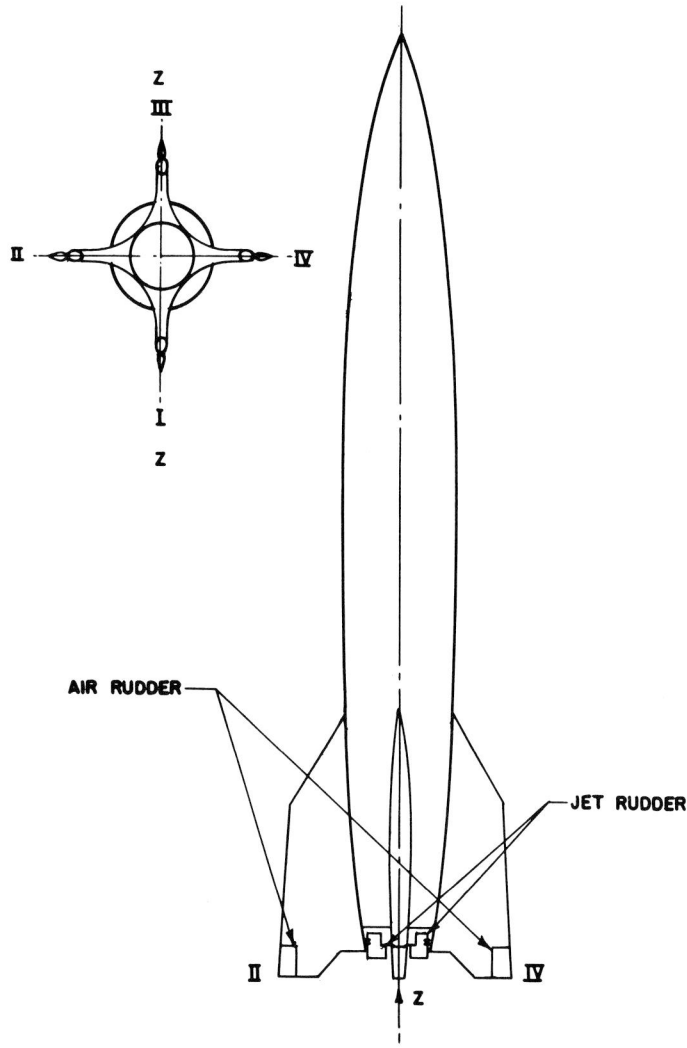

Location of the graphite jet vanes and fin rudders that guided the missile during powered flight. *Source: Summary Report of A-4 Control and Stability.*

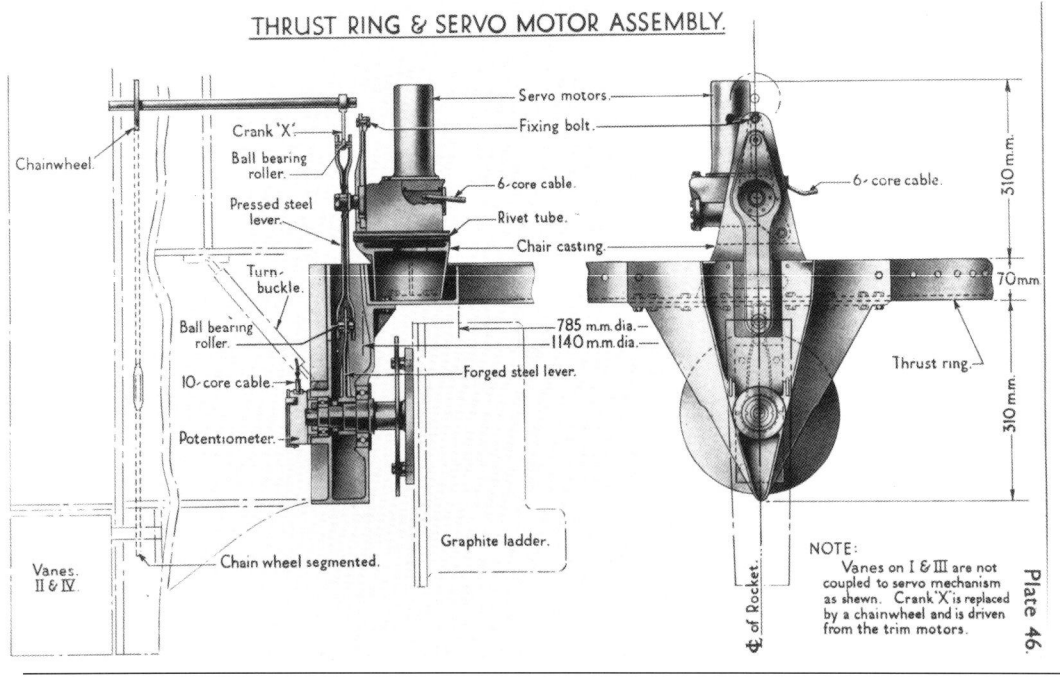

Details of the system of jet vanes and rudders that guided the V-2 during powered flight. *Source: Report on Operation Backfire.*

Chapter 8: V-2, From the Inside

oven for an hour. To guard against absorption of moisture, the finished vanes were coated with lacquer. As a final quality-control measure, the finished vanes were strength-tested in a machine where hydraulic rams applied a force of 2,200 pounds against the top of each one. Four jet vanes were needed for each rocket, but they were packaged in sets of five to allow for breakage during assembly and handling.

The missile propulsion section comprised three major subassemblies: the combustion chamber; turbopump; and steam generator. As its name implies, the combustion chamber was where the propellants burned. The hot combustion gases exited through a nozzle at the chamber's base to produce the motor's thrust. A steam-powered turbopump forced the liquid oxygen and alcohol propellants into the combustion chamber. Steam for the turbopump came from the decomposition of hydrogen peroxide in the steam generator. Sodium permanganate was used as a catalyst to decompose the peroxide.

The combustion chamber consisted of the head and lower sections. Both sections were made from steel. The head contained the

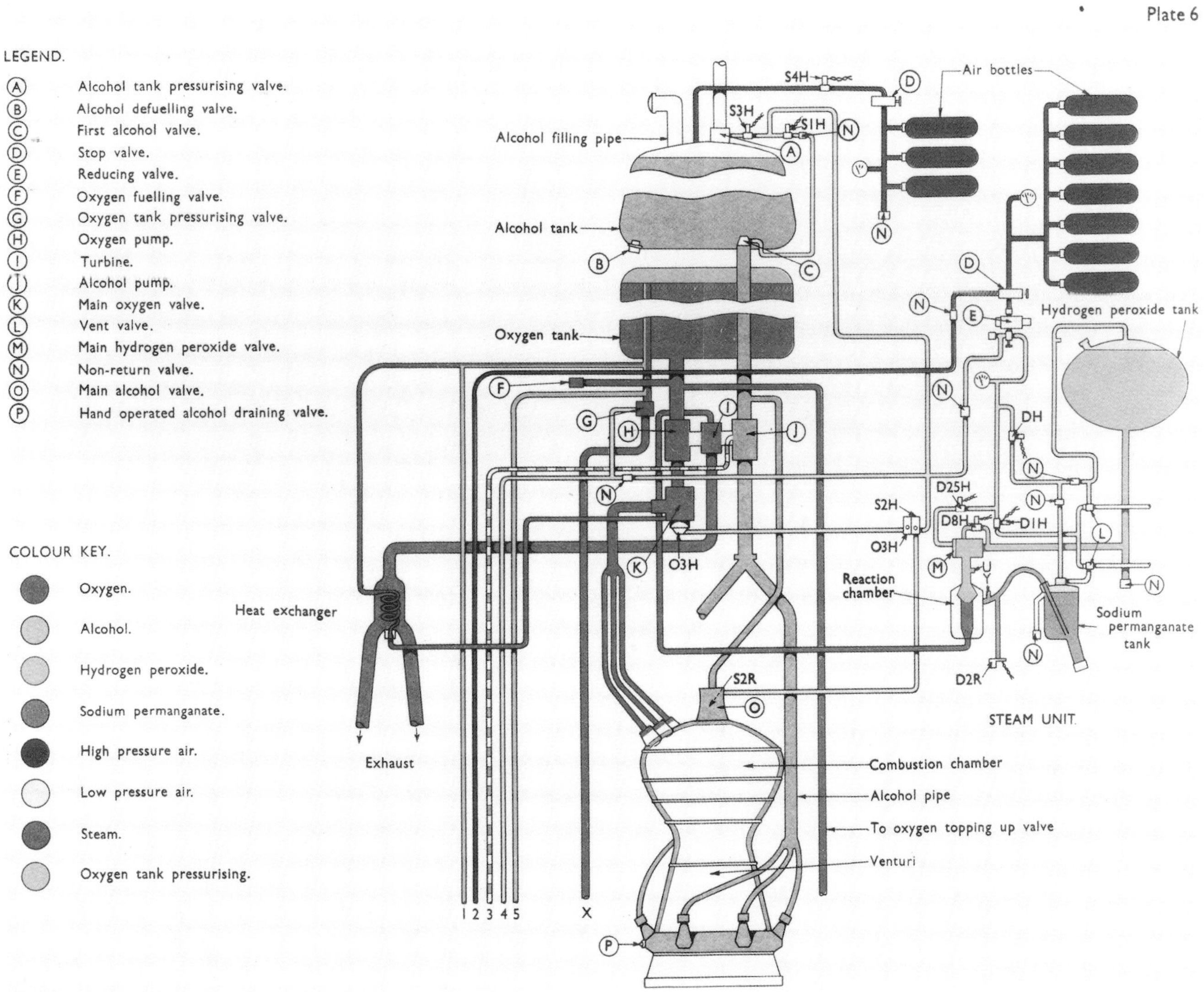

Graphite jet vane. *Source: Summary Report on A-4 Control and Stability.*

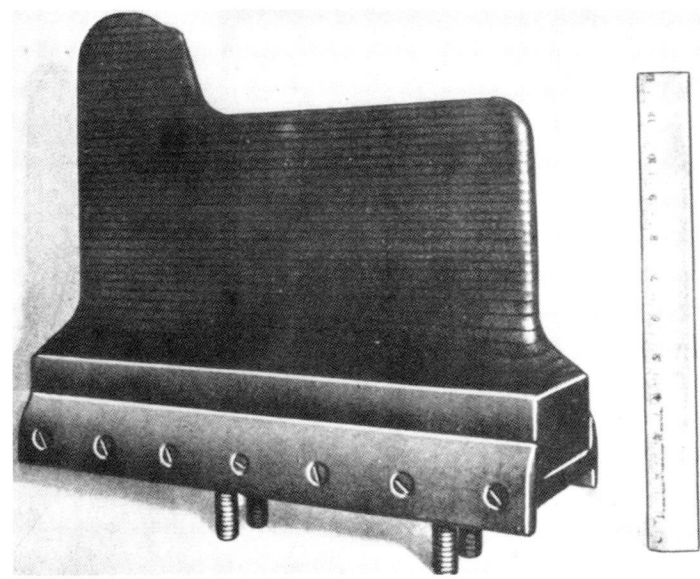

Propulsion system schematic. *Source: Report on Operation Backfire.*

18 propellant injectors, cup-like assemblies that had thousands of small nozzles that atomized the propellants, and the main alcohol inlet. All together, there were 2,160 nozzles for the liquid oxygen and 1,224 for the alcohol. The combustion chamber was a double-walled assembly. During firing, alcohol circulated between the walls to cool the chamber and prevent burn-throughs.

The three subassemblies of the propulsion section were mounted on a tubular steel frame that attached to the head of the combustion chamber at four points. The propulsion section weighed 2,050 pounds.

In early liquid-fuel rockets like the A-2 and A-3, propellants were forced into the combustion chamber by gas pressure. For a large rocket like the V-2, such an arrangement would have been too heavy. Therefore, the V-2 used a turbopump to feed propellants into the combustion chamber. A steam turbine that generated 460 horsepower with a rotation of 3,800 revolutions per minute powered two pumps. One pump was for the liquid oxygen, the other for the alcohol. Since liquid oxygen reacts violently with organic materials, oil-less bearings had to be used for the oxygen pump. If the turbine reached a speed of 4,500 revolutions per minute, the steam supply was interrupted to prevent damage to the rocket.

During the first few seconds after ignition the propellants were gravity fed to the combustion chamber, the so-called "preliminary stage." During the preliminary stage, the engine generated about 5,000 pounds of thrust. Although this was not enough to make the rocket lift off, it allowed the firing control officer to observe the function of the motor. Once he was satisfied that all was functioning normally he ordered the "main stage," or full thrust. Hydrogen peroxide was allowed to enter the steam generator, and the turbopump quickly came up to speed. Fuel consumption increased to 38 kilograms of liquid oxygen and 35 kilograms of alcohol per second, and the thrust built up to 56,600 pounds.

For those on the ground, particularly those who had never seen a large rocket motor in operation, a V-2 launch was an awesome

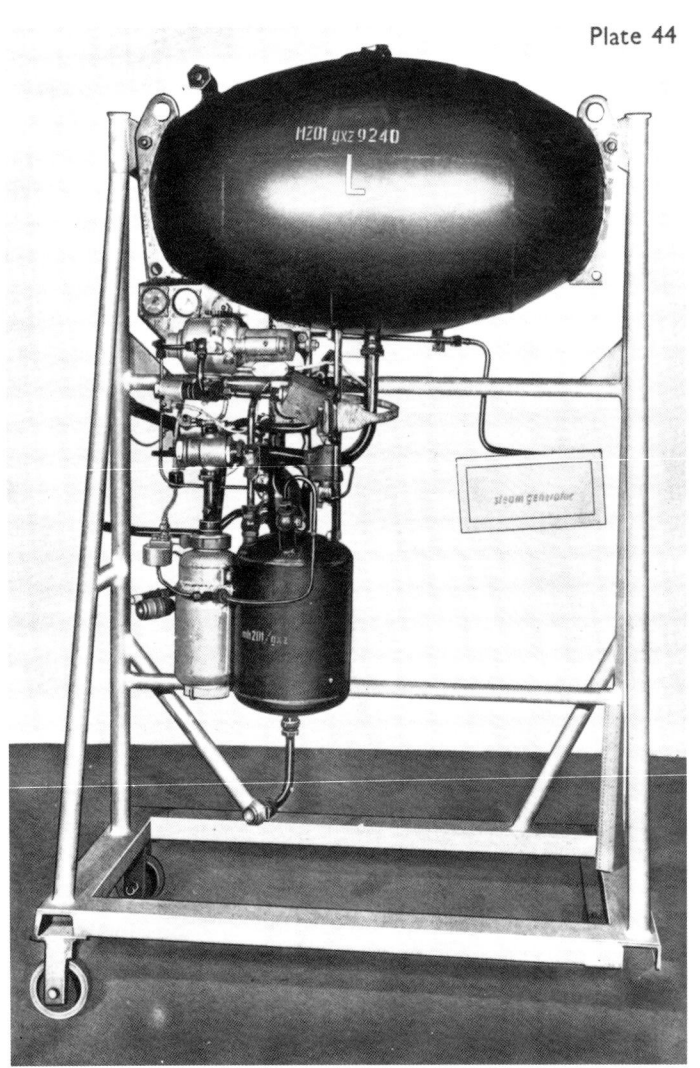

Steam generator unit. The egg shaped tank at the top contained the highly volatile and sensitive hydrogen peroxide; the smaller cylindrical sodium permanganate tank is visible near the base of the unit.

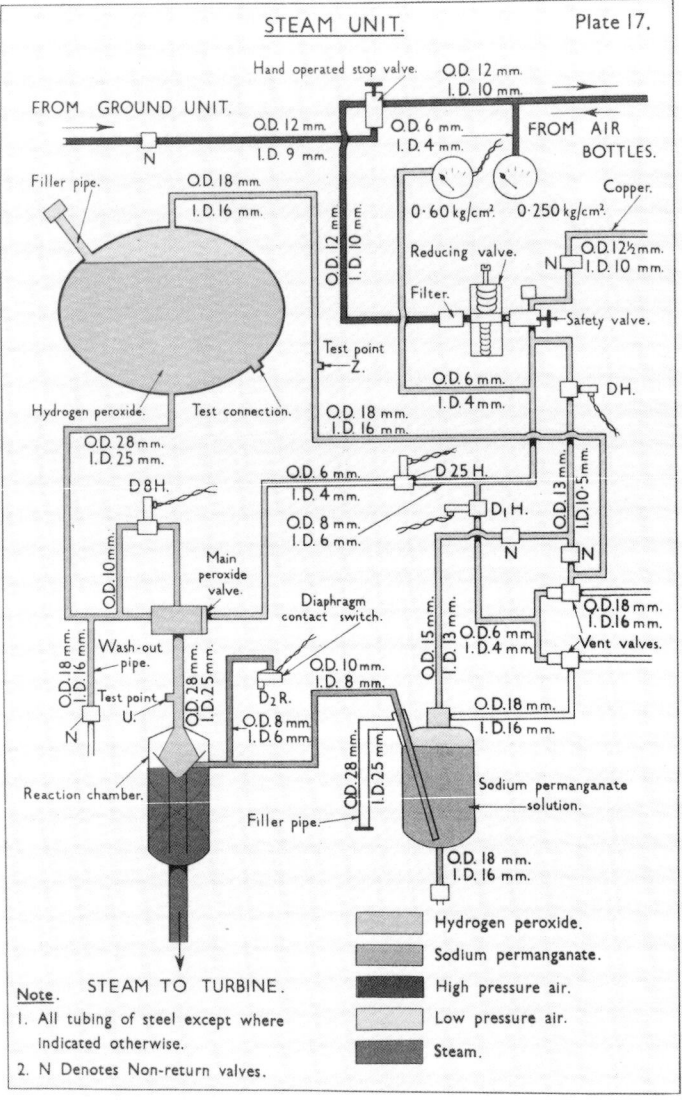

Steam unit schematic. *Source: Report on Operation Backfire.*

Chapter 8: V-2, From the Inside

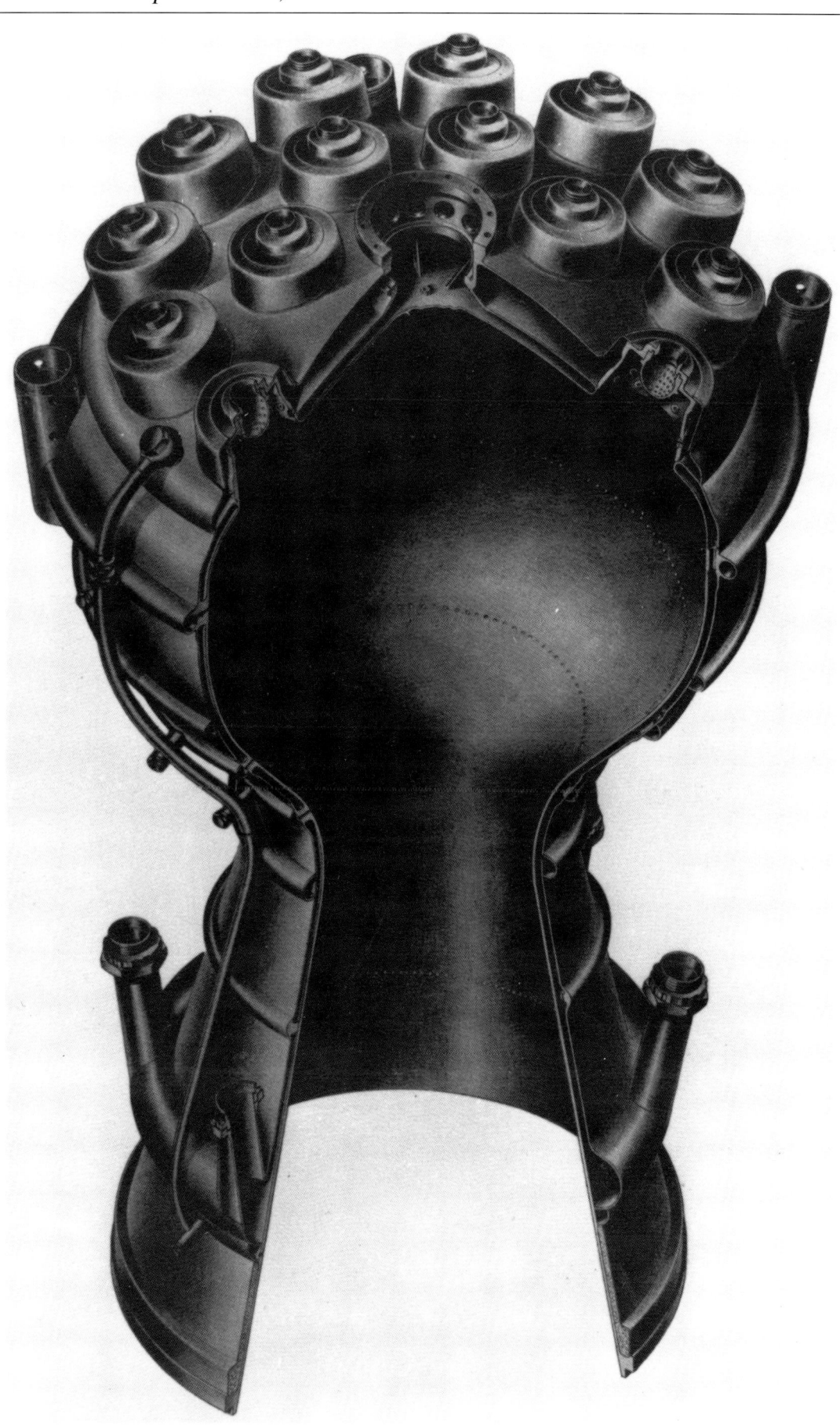

Combustion chamber cutaway. This cutaway drawing of the combustion chamber clearly shows the double-wall construction, cooling holes in the nozzle throat area, and burner cup arrangement.

spectacle. Soon, residents in Holland and France would witness this spectacle many times.

A-4 Statistics

Overall Length	14,036 mm
Length of Warhead	2,010 mm
Length of Control Compartment	1410 mm
Length of Midsection	6,215 mm
Width of Gap Between Midsection and Tail	5 mm
Length of Tail	4401 mm
Length of Fins	3,935 mm
Body Diameter	1,651 mm
Fin Span	3,564 mm
Empty Weight	4,000 kg
Loaded Weight	12,700 – 12,900 kg
Alcohol Fuel Consumption (Main Stage)	58 kg per second*
Liquid Oxygen Oxidizer Consumption (Main Stage)	72 kg per second*
Pressure in Combustion Chamber	14.5 atmospheres
Temperature in Combustion Chamber	2,000° Celsius
Range	300 km*
Altitude at Top of Trajectory	80 km*
Altitude at End of Combustion	28 km*
Combustion Time	60 – 63 seconds
Velocity at End of Combustion	1,500 meters per second*
Time of Flight	320 seconds*
Velocity at Impact	800 meters per second*
Initial Acceleration	1 g (32 feet per second per second)
Acceleration at burnout	6 g

*approximate values – varied slightly from round to round.
Source: Das Gerät A-4.

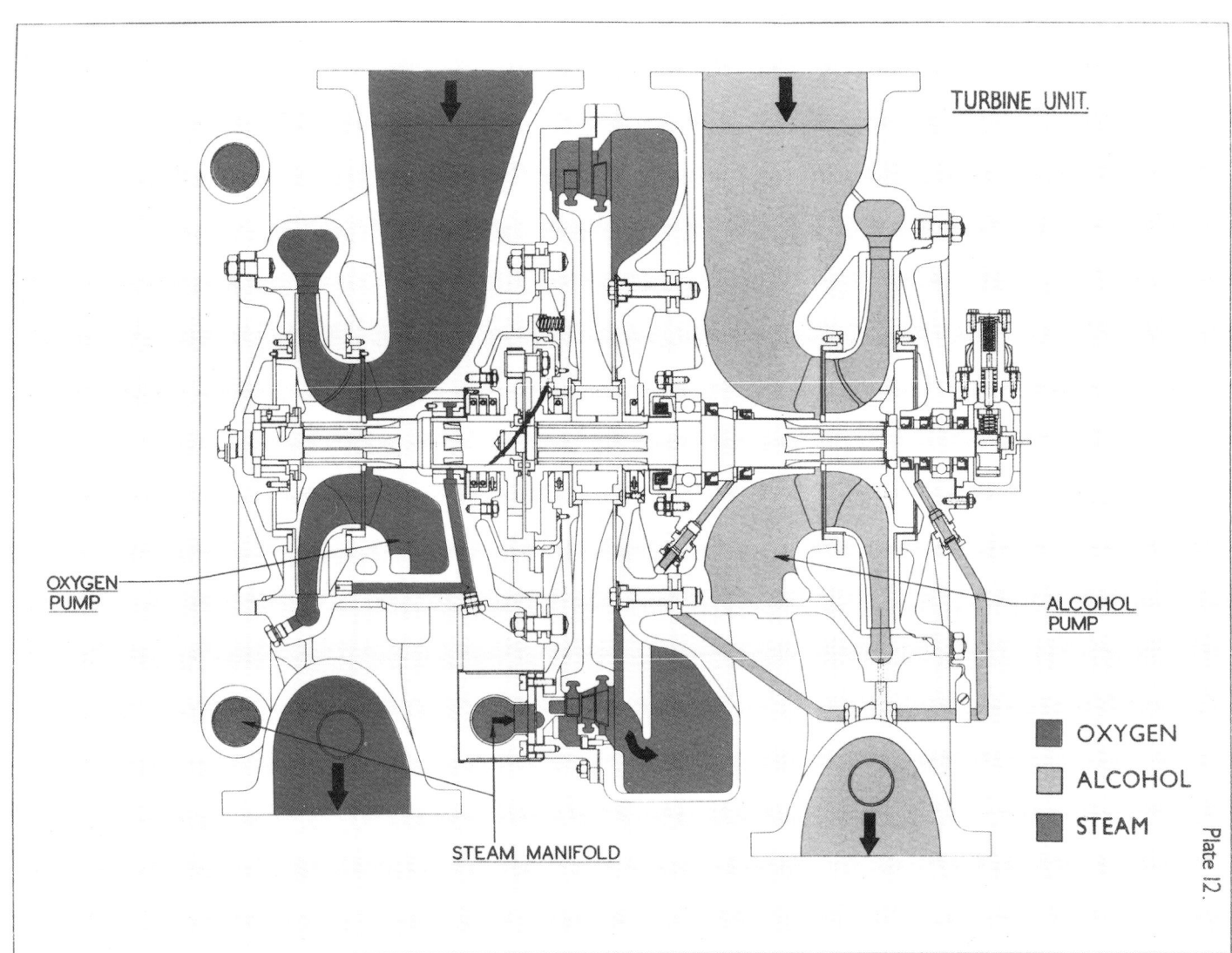

Propellant turbopump cutaway. Source: Report on Operation Backfire.

9

The V-2 at War

Under Kammler's improvised headquarters, the V-2 field command comprised two groups: Group North and Group South. Group North consisted of the 1st and 2nd Batteries of the 485th Battalion, while Group South comprised the 2nd and 3rd Batteries of the 836th and the 444th *Lehr und Versuchs Batterie*. The 444th was ordered to Fraiture, near Vielsalm, in the Ardennes.

The distinction of firing the first shot of the V-2 offensive went to the 444th, which fired two rockets towards Paris on 6 September. Both failed. The battalion moved to a site 10 miles south of Houffalize and launched another missile towards Paris on the morning of 8 September. This rocket impacted near the Port d'Italie in Paris, 180 miles from the launch pad.

V-2 operations against England began from launch sites in Holland. On 7 September, German personnel went door to door in a neighborhood in Wassenaar, a northern suburb of the Hague, and told the residents they had two hours to vacate their homes. They were told to leave everything behind, and to open all their windows and glass doors. The following evening, at around 6:45 PM local time, the 485th sent two missiles towards London. Their aiming point was 1,000 yards east of Waterloo Station. The first missile landed

In March 1945, towards the end of the V-2 campaign, a missile fell near the junction of Wanstead Road and Endsleigh Gardens in Ilford, England. Nine people were killed, eight houses destroyed, and another 16 so severely damaged they had to be demolished, along with another 149 damaged.

at Chiswick, in the northern suburbs of London. It left three people dead and 10 injured. The second missile hit 16 seconds later near Epping. There were no casualties from this missile.

Firings continued from the Hague-Wassenaar area until 17 September, when Group North withdrew to Burgsteinfurt because of the Allied airborne landings near Arnhem. Meanwhile, the 444th moved to the island of Walcheren and began a special operation against London under the personal supervision of Kammler. Assisted by the 91st Technical Artillery Battalion, they launched six rockets against the British capitol between 16 and 19 September. The Allied landing at Arnhem eventually forced them to move to Zwolle.

Withdrawal of both battalions meant there were no V-2 units within range of London, so Kammler ordered the 444th to bombard Norwich and Ipswich from positions near Stavorem, in Friesland. The 485th remained in Burgsteinfurt and fired on Louvain, Tournai, Masstricht, and Liége. This continued until 30 September, when the 2nd Battery of the 485th returned to the Hague and resumed operations against London, while the 1st Battery engaged targets on the European continent. Further south, the 836th moved to the Euskirchen area and launched missiles against Lille and Mions.

Many organizational problems appeared during the first weeks of the V-2 campaign, due to the lack of staff control within Kammler's improvised headquarters. All too often, rockets arrived at the firing points without warheads or spare parts. A shortage of gasoline trucks prompted the Division supply officer to use alcohol tankers to transport motor fuel. This ruined the tankers for carrying rocket fuel. To alleviate this situation, Kammler requested help from the *Oberkommando des Herres* (OKH, or High Command of the Army). On 30 September the OKH organized a divisional headquarters for Kammler. His division was called simply "Division zV." The "zV" stood for *"zur Vergultung"* (for Vengeance).

While the V-2 barrage is most frequently linked with London, an even greater number of missiles hit targets on the European continent, notably Antwerp. The continental campaign began slowly at first. Following the single V-2 that hit Paris on 8 September, no more rockets were fired against Continental targets until the 13th, when a rocket landed near Brussels. The next day can really be regarded as the start of the Continental campaign, when four more missiles hit Brussels. During the next month Liége, Lille, Brussels, and Ghent were hit. Liége, an important rail and communications center, bore the brunt of the attack, with an average of four incidents per day.

By this time the 836th had moved across the Rhine to continue bombarding continental targets. On 4 October Himmler ordered the cessation of firing on Paris. Eight days later Hitler directed that all firing be concentrated on London and Antwerp, Belgium. Antwerp had become the main port on the European continent for the influx of supplies and personnel for the Allies, and Hitler hoped to neutralize the port's effectiveness. The center of the city and its docks were selected as the aiming points for the missiles. Groups North and South redeployed to comply with the *Führer's* latest directive.

During October the 3rd Battery of the 485th completed its training at Heidekraut, and joined Group North in the Burgsteinfurt area. A new battalion, the SS *Werfer Batterie* 500, also joined Group North, and was deployed around Hellendoorn-Zwolle, in Holland. These units opened fire on Antwerp on 13 October. For the next two months an average of nine rockets hit the Belgian capital each day. The tempo intensified, reaching a peak of over 100 per week for three weeks beginning in mid-December.

On the German home front, and in the international media, the Nazi Propaganda Ministry made all sorts of exaggerated claims about the effects of the V-2. These claims increased with the Christmas time counteroffensive (Battle of the Bulge), in an effort to reassure the German populace that Allied attacks on targets in the Reich were being avenged. The Nazis reported:

"...up till now this 'V-2' has destroyed three Thames bridges in London. Despite official British measures to keep the damage secret, it is inevitable that the increasing amount of information on the damage done by the 'V-1' and the 'V-2' in London—which travelers describe as 'extremely extensive'—should seep through."

The German communiqué, which was filed through neutral Madrid, further claimed:

"...the Houses of Parliament have been damaged extensively. There is not a building standing within 500 meters of Leicester Square. Piccadilly Circus has also been devastated. The Tower has suffered considerable damage from blast. The number of 'V-2' bombs which have exploded right in the center of London is considerable."

While the V-2 was causing a great deal of damage in London, it was not nearly as extensive as the Nazis claimed. When the first V-2s hit London, the British Government withheld information about the nature of the weapon from the populace. Explosions were attributed to exploding "gasworks," or similar phenomena. The V-2 struck without warning. There would suddenly be a tremendous explosion (that left a large crater) and tremor, followed by a distinctive double report caused by the sonic boom that trailed the missile. On November 11, 1944, Prime Minister Churchill formally announced to the British public that London was under attack by the V-2. London newspapers on 10 November carried stories about the new weapon, and one even had a fairly accurate illustration of the missile.

Throughout the V-weapon campaign in London, the V-2 caused more than double the number of deaths per missile than the V-1. The V-1 caused an average of 2.2 deaths per bomb; for the V-2 the average was 5.3. Several factors contributed to the higher casualty rate. One of the major factors was the lack of warning. There were reports of people on the ground seeing a dull red glow just before missiles hit after dark, but during daylight hours (which is when the overwhelming majority of incidents occurred) there was no indication a missile was enroute before it exploded.

In contrast, the relatively slow V-1 was very noisy, and gave ample warning of its approach. Just before it began its dive to the ground, the cruise missile's engine would quit, too. The V-1 had a small propeller on its nose that turned as the missile flew through

Chapter 9: The V-2 at War

the air. After the propeller had turned a certain number of revolutions, it had traveled the prescribed distance and was over the target. The elevator locked in the down position, sending it crashing to the ground. This action was so violent that it disrupted the fuel flow to the engine and made it stop about 8-10 seconds before the missile hit. The sudden silence warned those in the area, and allowed them to assume a protective posture; even lying along a curb would frequently offer sufficient protection to survive a relatively close explosion. This defect in the V-1 design probably saved hundreds of lives.

Another factor contributing to the V-2's greater lethality was the intensity of damage it caused. The average radius of damage where houses were 75-100 percent destroyed was nearly identical for both the V-1 and V-2 (72 versus 76 feet), but within that circle damage from the V-2 was more complete. That is, the rubble was more completely pulverized by the V-2, so those who were not killed by the blast were frequently buried and suffocated. The zone of secondary damage (50-75 percent destruction) was again nearly identical for both missiles, but the radius for lesser damage was significantly greater for the V-2. In this zone broken glass caused many casualties.

There were also several incidents where a single V-2 caused more than 100 fatalities. One such incident took place in November at the New Cross Shopping Center. A V-2 struck just before noon, when Christmas shoppers crowded the Woolworth's department store. It took rescue workers nearly a week to remove 168 bodies from the ruins. There were also 108 seriously injured. Another midday incident in March 1945, near the end of the V-2 campaign, killed 115, and left 123 seriously injured. This occurred at Smith Market and Farrington, where a large number of women were in line to buy groceries. As with the New Cross Shopping Center incident, it took several days to exhume all the bodies from the rubble.

On January 1, 1945, the 836th became the 901st Motorized Regiment, and the 485th became the 902nd. Both were part of Kammler's Division zV. SS *Werfer Batterie* 500 became a battalion, and was assigned to the 902nd, giving the Regiment four firing battalions. The 901st comprised three battalions.

In a similar vein, the V-1 field command also underwent several reorganizations. The LXV Armee Korps became the XXX Armee Korps on October 19, 1944. General von Treskow replaced Heinemann as Corps Commander on 26 October, but he held this post for less than three weeks. General Kleffell became XXX Corps Commander on 15 November, and held this post until January. On January 28, 1945, the XXX Corps became the 5th Flak Division, and was placed under Kammler's control. With the addition of the V-1 field command, the Division zV became the Armee Korps zV. The V-1 and V-2 then remained under Kammler's authority until the end of March.

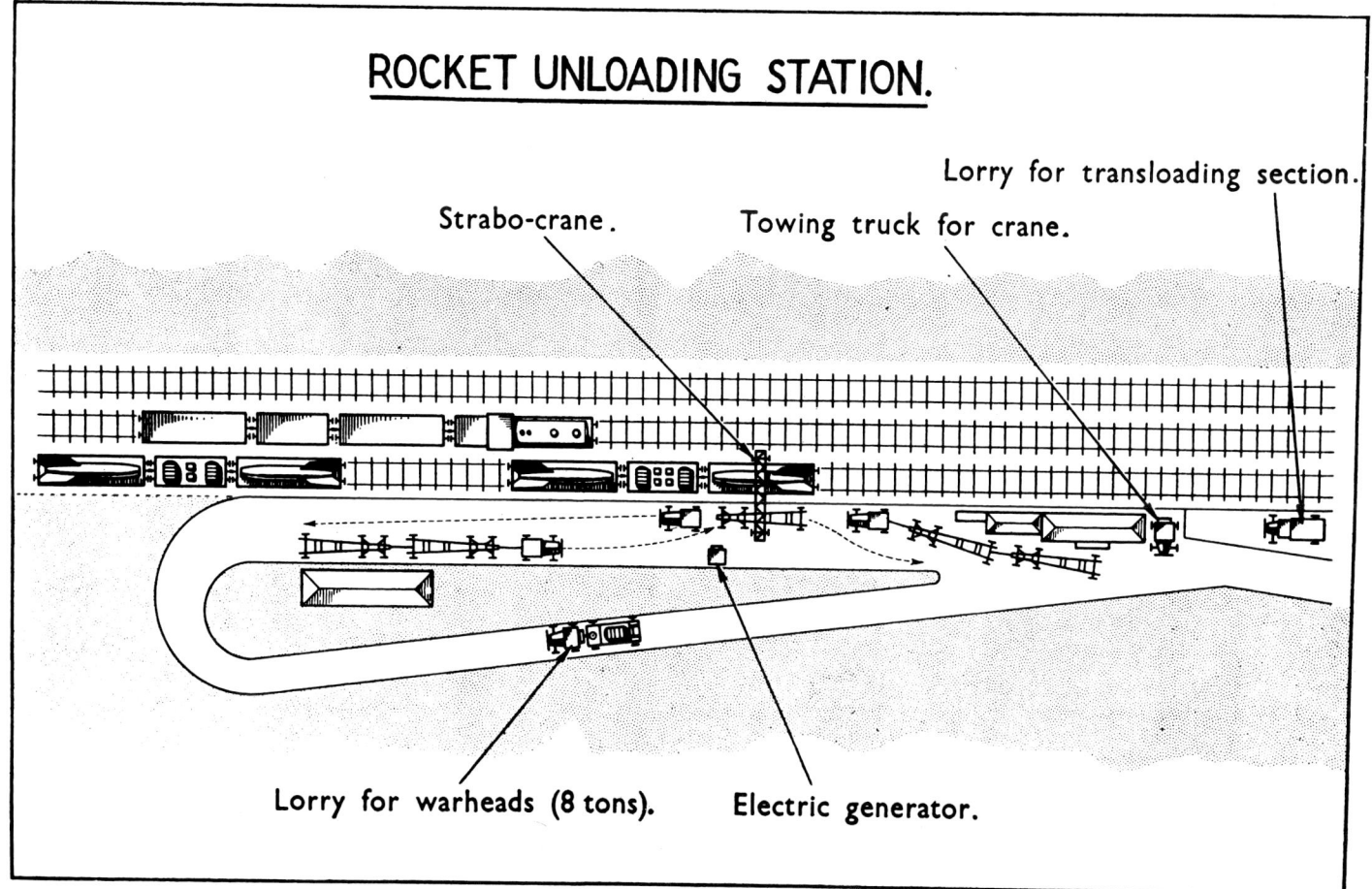

Layout of the unloading station. Source: *Report on Operation Backfire*.

After the reorganization of the V-2 units, antiaircraft batteries, an engineer battalion, a supply division, and three protective battalions were added to the Division zV for support. Each regiment, in addition to the missile batteries, contained a headquarters section, a workshop platoon for motor vehicles, two light antiaircraft batteries, a medical section, a supply section, an engineer battalion, and a maintenance company for motor fuel and oil transport. The V-2 Battery comprised a Headquarters, Headquarters Troop, Launching Troop, Fuel and Rocket Troop, and a Technical Troop. These units contained small arms, communications gear, trucks, and other equipment you would expect to find in any typical military unit. Due to the unique nature of the V-2, however, there were also such specialized items as rocket transporters, fuel and oxidizer trailers, and launch pads.

Experience with V-2s in the field showed the longer a rocket was in storage, the greater the likelihood it would malfunction. Therefore, the Germans adopted a scheme they nicknamed the "Hot Cakes" system, where the rockets were transported from the Mittelwerk to the field units as quickly as possible. Trains went from Mittelwerk to a special outfitting station, where the warheads, jet vanes, fuses, and containers of sodium permanganate were loaded. Each train carried 20 missiles. The trains were organized in to three-car sections, each one carrying two missiles. Cars one and three of each section held the missiles; the middle car was used for the warheads, sodium permanganate, and graphite jet vanes. Camouflage covers over the cars concealed their cargo.

The Technical Troop was in charge of unloading the missiles from the trains and preparing them for flight. There were specific platoons or sections assigned to each individual task needed to test and prepare a rocket. Interestingly, once the Technical Troop platoon responsible for unloading the rockets had them off the rail cars, they turned them over to the Fuel and Rocket Troop, who moved the missiles to the Technical Troop platoons, who prepared them for flight.

The Fuel and Rocket Troop contained three platoons, each one with specific responsibilities for a particular propellant or task. The first platoon transported liquid oxygen from the rail site to the launching area; the second platoon handled the alcohol fuel and hydrogen peroxide needed for the turbopump. The third platoon moved the rockets and warheads from the railhead to the Technical Troop area. This platoon also handled the sodium permanganate that was used with the hydrogen peroxide in the turbopump.

There were three firing platoons in the Launching Troop, each with one launch pad. A launching platoon was expected to fire two or three missiles per day.

Kammler selected the overall regimental areas. Each Regiment Commander then assigned an area to each of his three firing batteries. The Battery Commanders then selected individual firing sites.

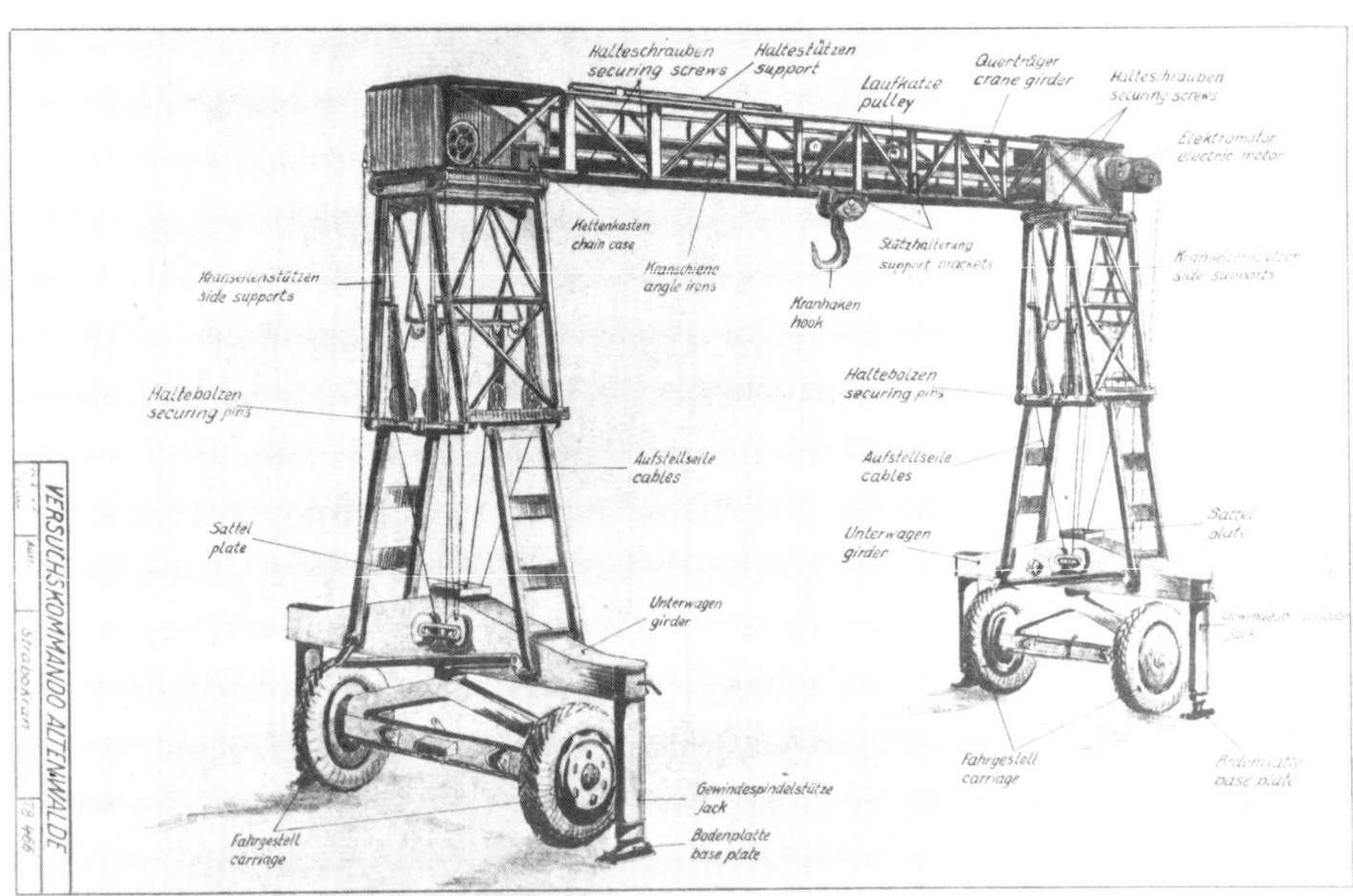

Technical Troop personnel used a portable crane, called the Strabo Crane, to unload rockets from the rail cars. Source: *Die Fernrakete*.

Chapter 9: The V-2 at War

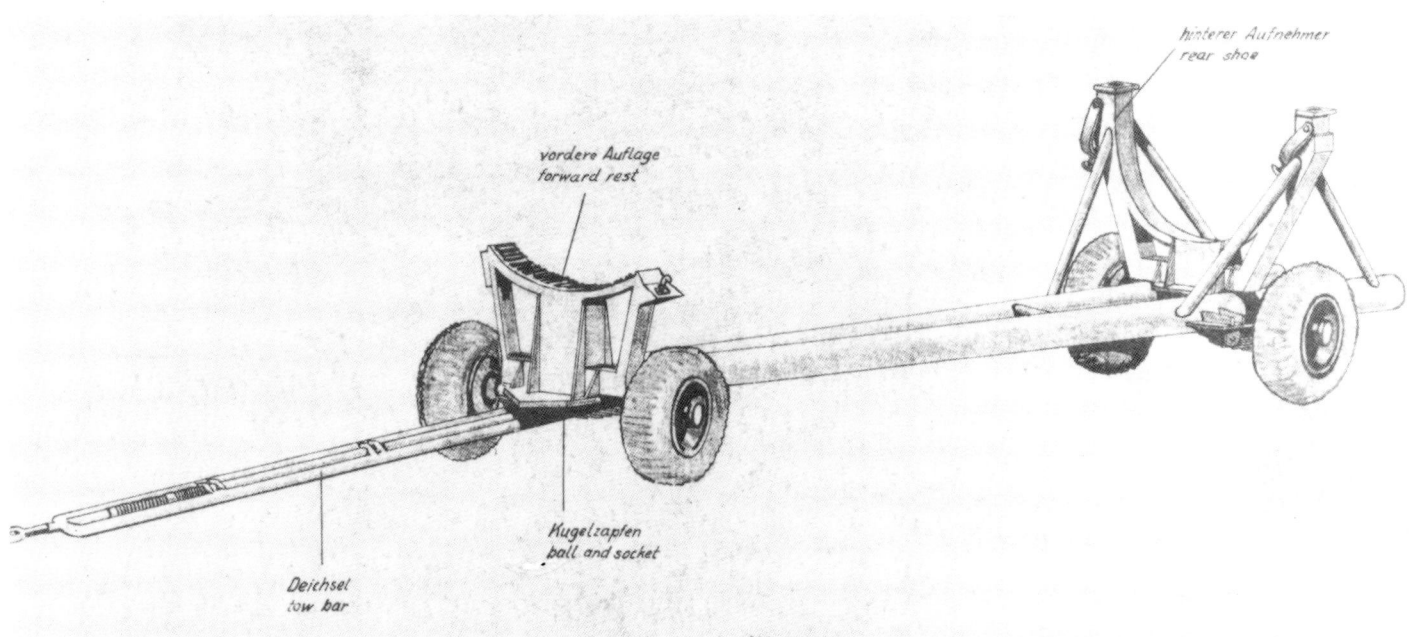

While the missile underwent preparation in the Technical Troop area, it was transported on a lightweight trailer called the Vidalwagen. *Source: Die Fernrakete.*

In the field, the launch sites were normally about 1,000 yards apart. Selection criteria for launch sites included proximity to railroads, quality and type of roads, and concealment. Pine woods were preferred, as the coniferous trees made aerial detection more difficult, and acted as a wind screen for the upright rockets.

The Railway Transport Officer, Regimental Quartermaster, Battery Quartermaster, and members of the Technical and Fuel and Rocket Troops met the trains at the unloading point. After the cargo was checked against the shipping voucher, the rockets were unloaded using a portable crane called the Strabo Crane. The Strabo Crane, which could lift 16 tons, consisted of a horizontal girder that rested on two scissor supports that were raised by an electric motor. A block and tackle traversed the length of the horizontal beam to lift the rockets off the cars and place them on trailers called Vidalwagens.

The Vidalwagen was a lightweight trailer used to transport a rocket from the railhead through the Technical Troop area. It was just over 46 feet long, and was used strictly to move the rockets in a horizontal position.

Technical Troops were situated between the unloading point and the Launching Troop. As with the selection of a launch site, overhead concealment was a factor in the dispersal of the unit. Concealment for the Technical Troop was particularly important, since they needed to erect several large tents, and required parking areas for missiles and vehicles. It also helped if the Technical Troop was situated along a road relatively close to the Launch Troop, so the missiles could be fired as soon as possible after their checkout.

After arrival in the Technical Troop area, the Testing Section checked the missile. Their checks included the propulsion unit, steering mechanism, alcohol tank pressurization, wiring, and switchover from ground to internal power. These checks usually took about three hours. If the Testing Section found any problems with the

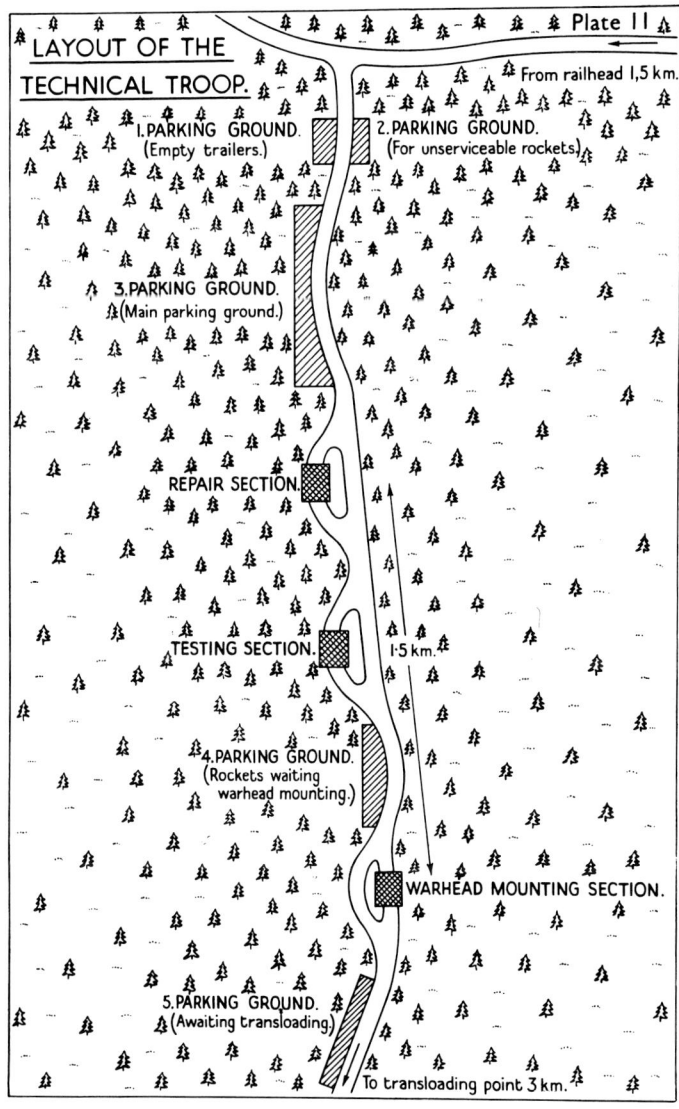

Layout of the Technical Troop area. *Source: Report on Operation Backfire.*

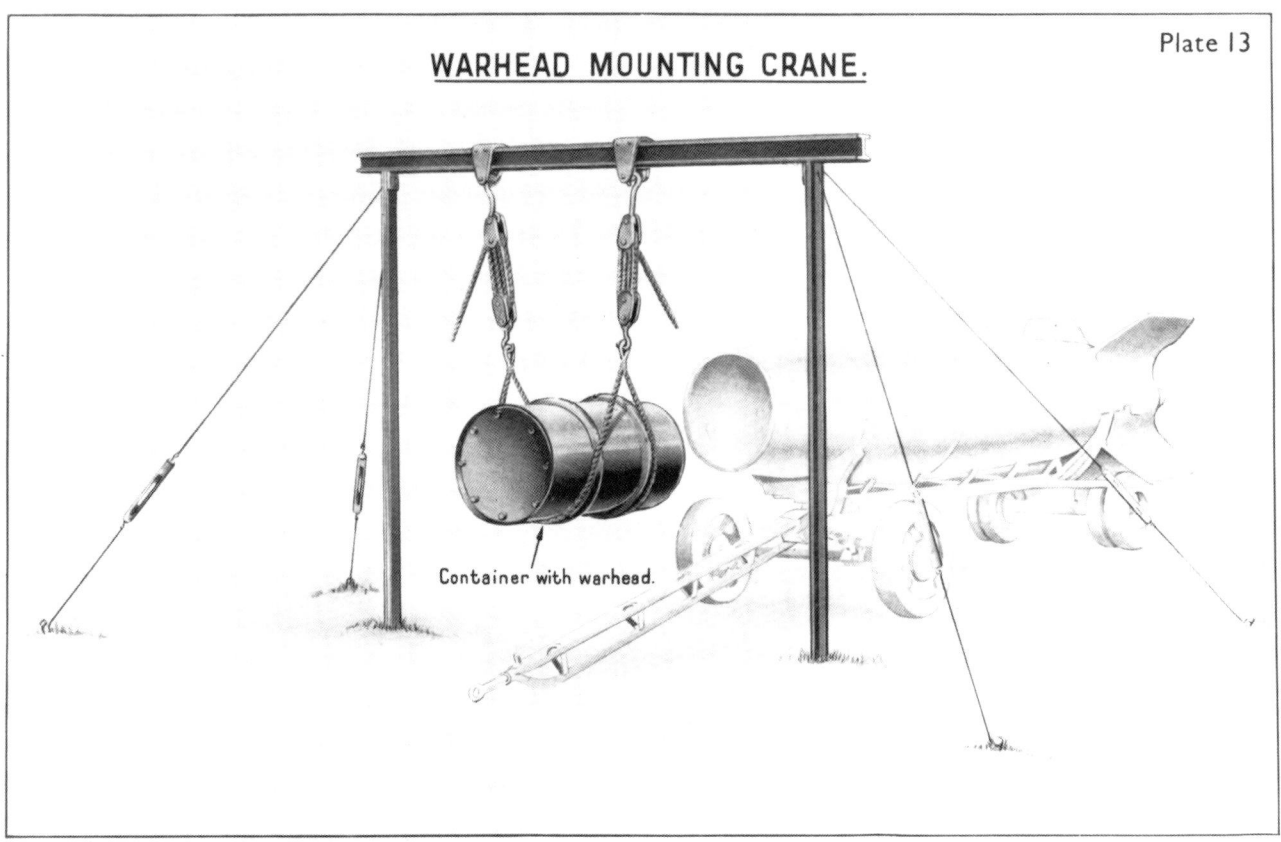

One of the last steps in preparing a V-2 was the attachment of the warhead and fuses. *Source: Report on Operation Backfire.*

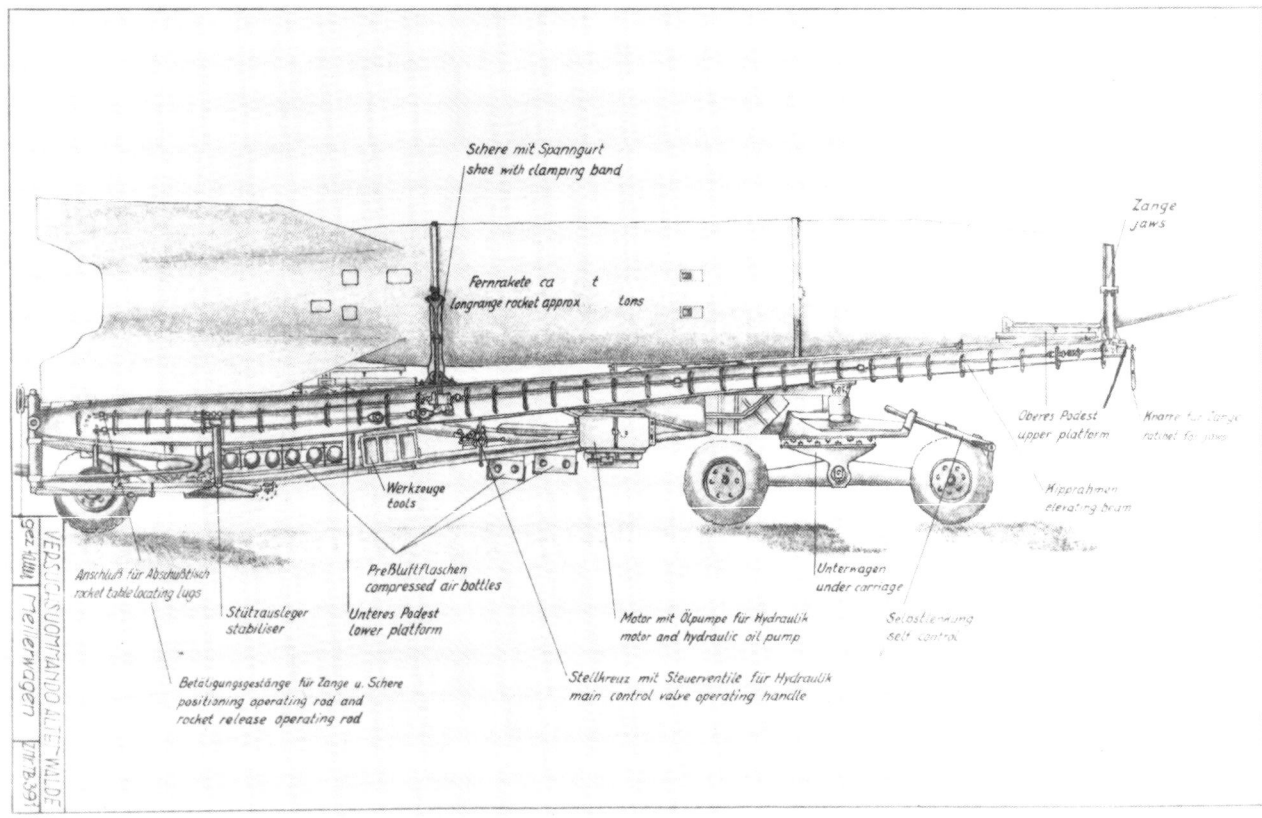

With the warhead attached, the Technical Troop used another Strabo Crane to transfer the missile from the Vidalwagen to the Meilerwagen, a combination transporter and service gantry. *Source: Die Fernrakete.*

Chapter 9: The V-2 at War

V-2 on the Meilerwagen. *Source: Mitchell R. Sharpe.*

Once the rocket reached the firing point, the Meilerwagen was used to raise it to a vertical position. *Source: Mitchell R. Sharpe.*

With the rocket in a vertical position, the Meilerwagen served as the gantry and servicing structure. *Source: Mitchell R. Sharpe.*

Chapter 9: The V-2 at War

missile, it was sent to the Repair and Tail Removal Section. This section performed only relatively minor repairs, like replacement of valves or servo motors. If a rocket needed extensive work, it was sent to the field workshop (*Feldspeicher*), or returned to the Mittelwerk.

Upon completion of the testing and any necessary repairs, the missile was taken to the Warhead Mounting Section. The warhead, still in its shipping container, was lifted into position with a block and tackle and attached to the rocket, which was still on a Vidalwagen. Once the warhead was attached to the rocket, the exploder tube was filled and fuses installed.

The next stop for the rocket was the transfer point where, using another Strabo Crane, it was transferred from the Vidalwagen to a Meilerwagen. The Meilerwagen was used to transport the missile to the firing point and erect it on the launch pad. Once the missile was in the vertical position, the Meilerwagen served as the servicing gantry. There were four main parts to the Meilerwagen: chassis; lift frame; hydraulic lift; and camouflage cover.

The lift frame attached to the chassis by a pair of trunnions. Two clamping collars on the frame supported and secured the rocket.

One collar held the rocket between the tail unit and mid section; the other was smaller, and encircled the missile at the base of the warhead. Three (later reduced to two) movable platforms attached to the lift frame to provide access to the rocket once it was in a vertical position. Rungs welded to the right side girder of the lift frame formed a ladder to reach the platforms. Plumbing for fueling the missile was contained in the lift frame.

Two hydraulic cylinders elevated the lift frame. Oil for the cylinders was stored in a cylindrical tank at the forward end of the chassis. A high-pressure oil pump, driven by either a 14-horsepower gasoline engine or electric motor, powered the cylinders.

Once the missile was on the Meilerwagen, a pipe framework and camouflage cover disguised the missile. Sockets were provided on the trailer for attaching the frame. Metal cases were also placed over the air rudders to protect them from damage while the missile was being moved.

The Launching Troop Rocket Supply and Accessories Column moved the rockets to the launch site, which had been prepared by the firing platoon. By the time the rocket arrived, the fire control vehicle (*Feuerleitpanzer*), launch pad, and electrical cables were already in place.

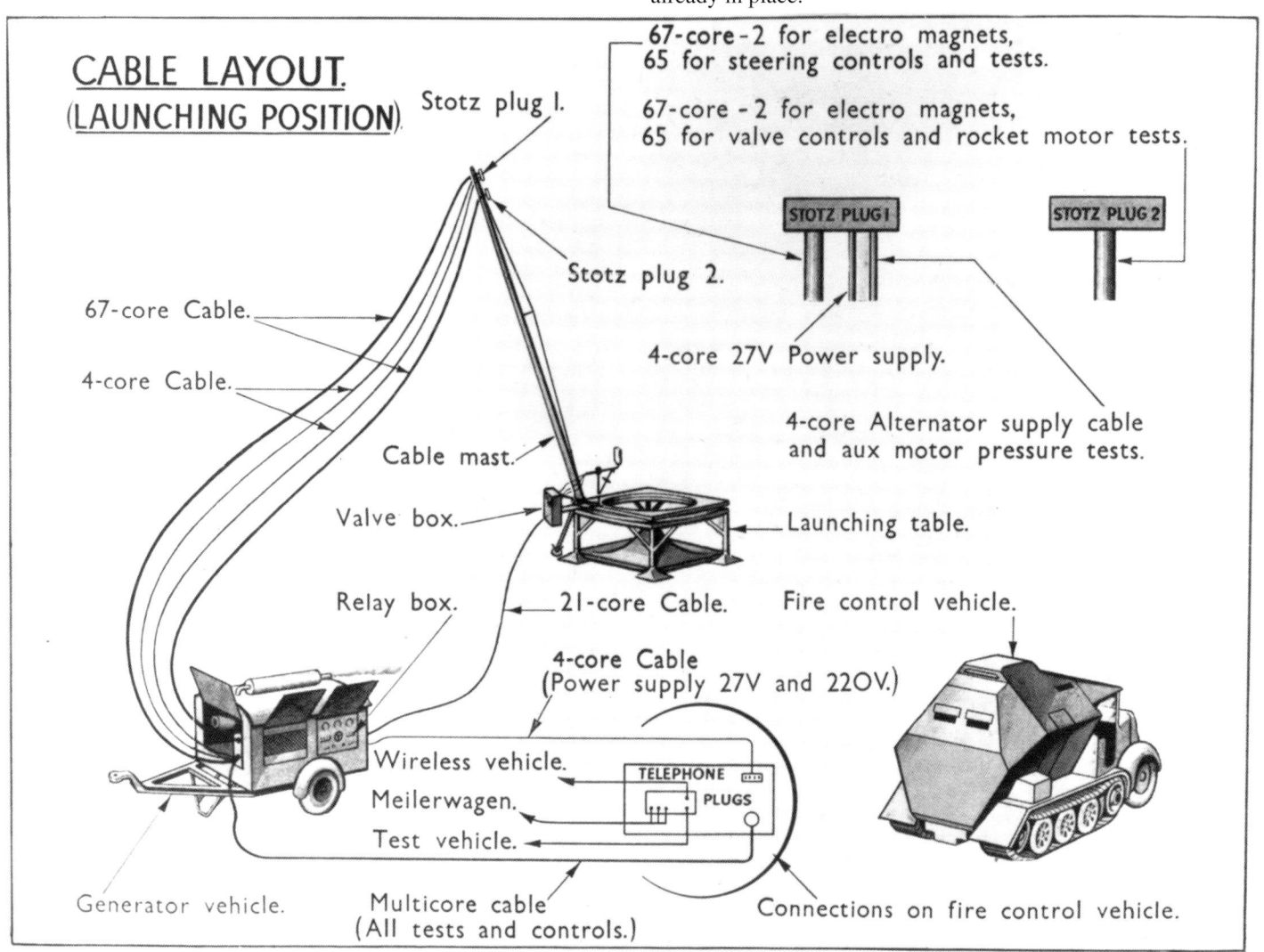

Cable layout in the launch area. *Source: Report on Operation Backfire.*

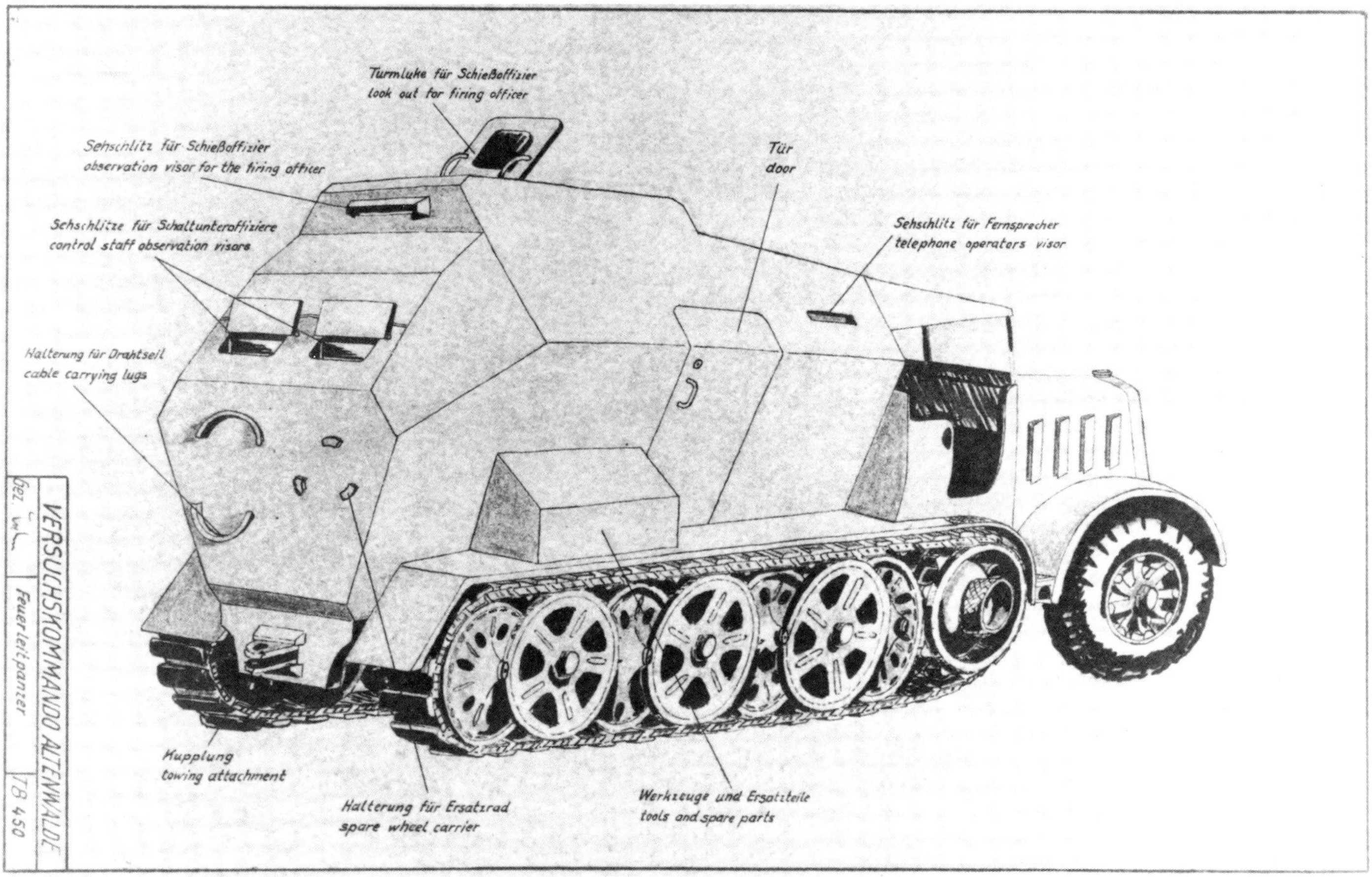

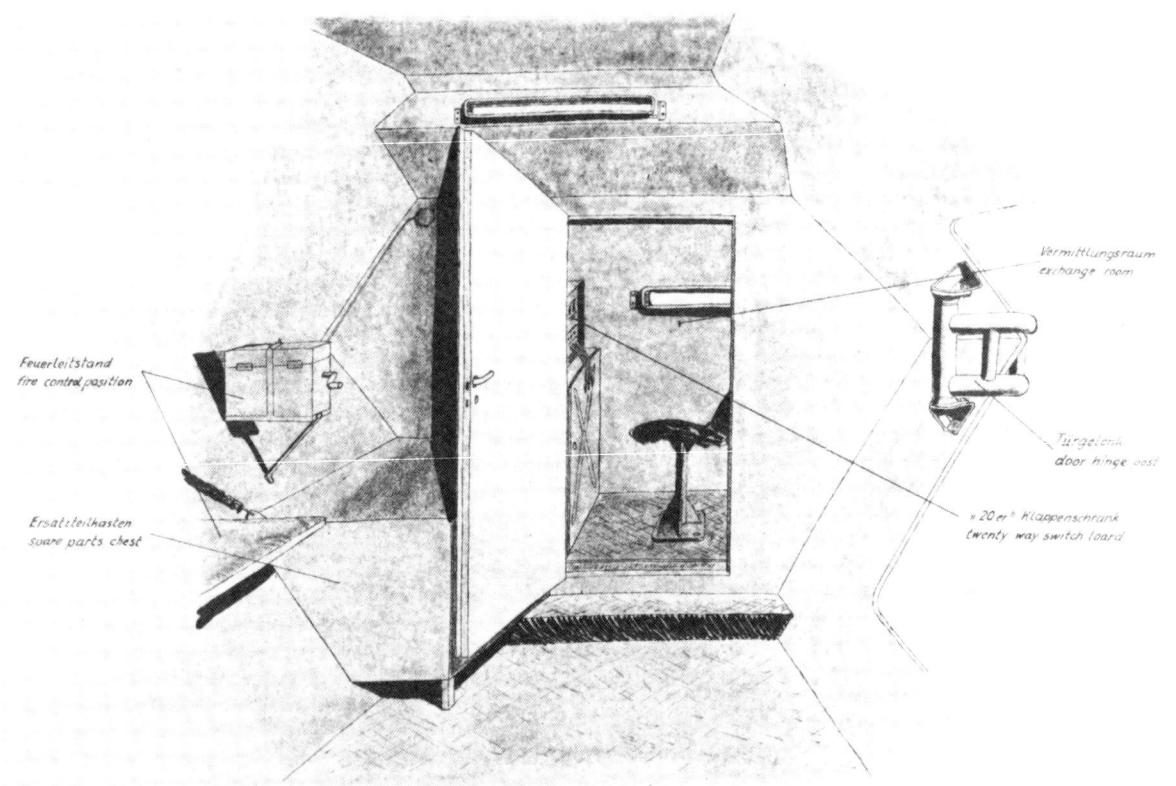

The fire control vehicle, or Feuerleitpanzer. The vehicle had compartments for the firing control panels and telephone switchboard. *Source: Die Fernrakete.*

Forward view inside the Feuerleitpanzer—the telephone switchboard compartment. *Source: Die Fernrakete.*

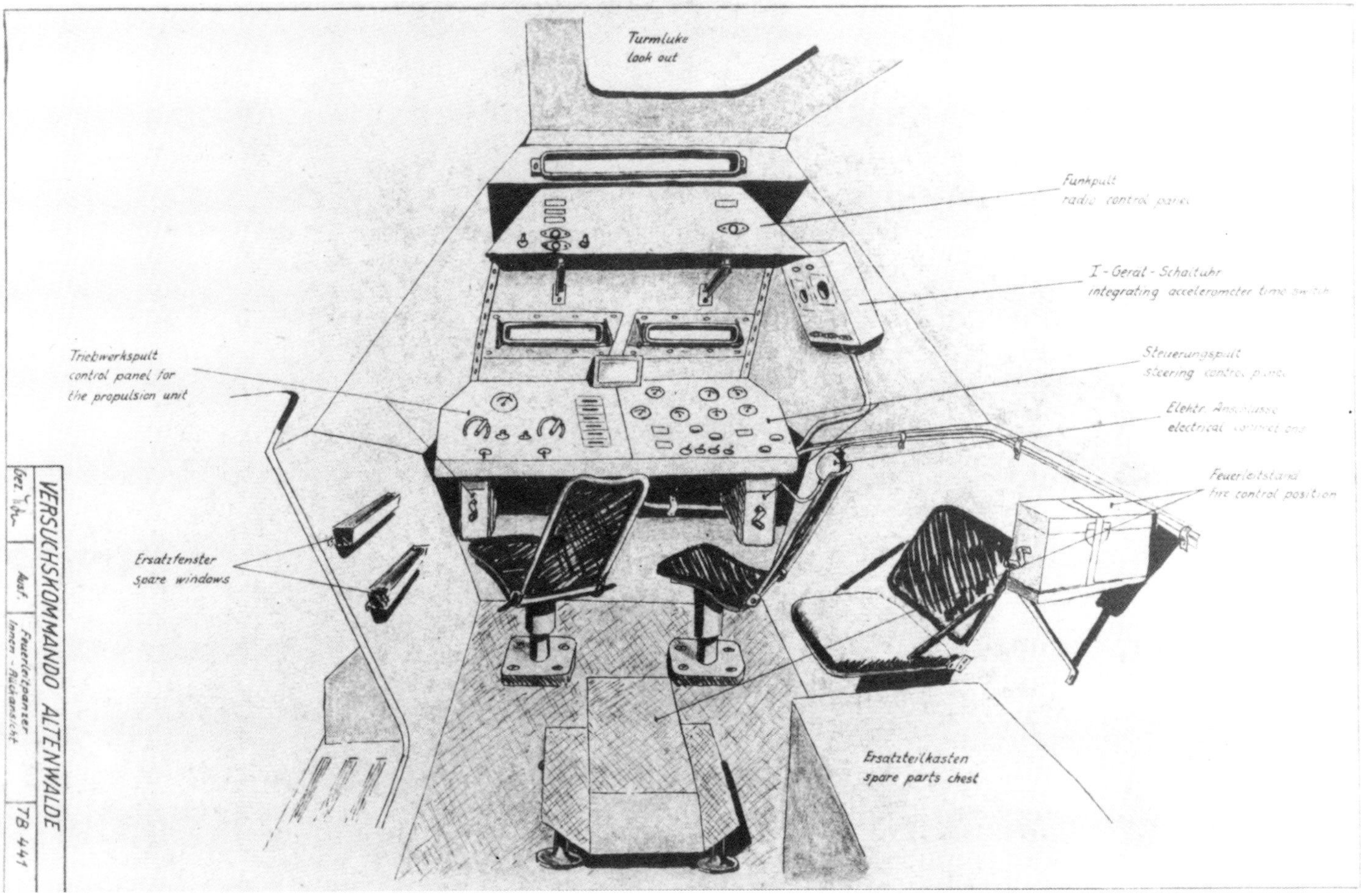

Chapter 9: The V-2 at War

The launch control and test positions in the Feuerleitpanzer. Source: *Die Fernrakete*.

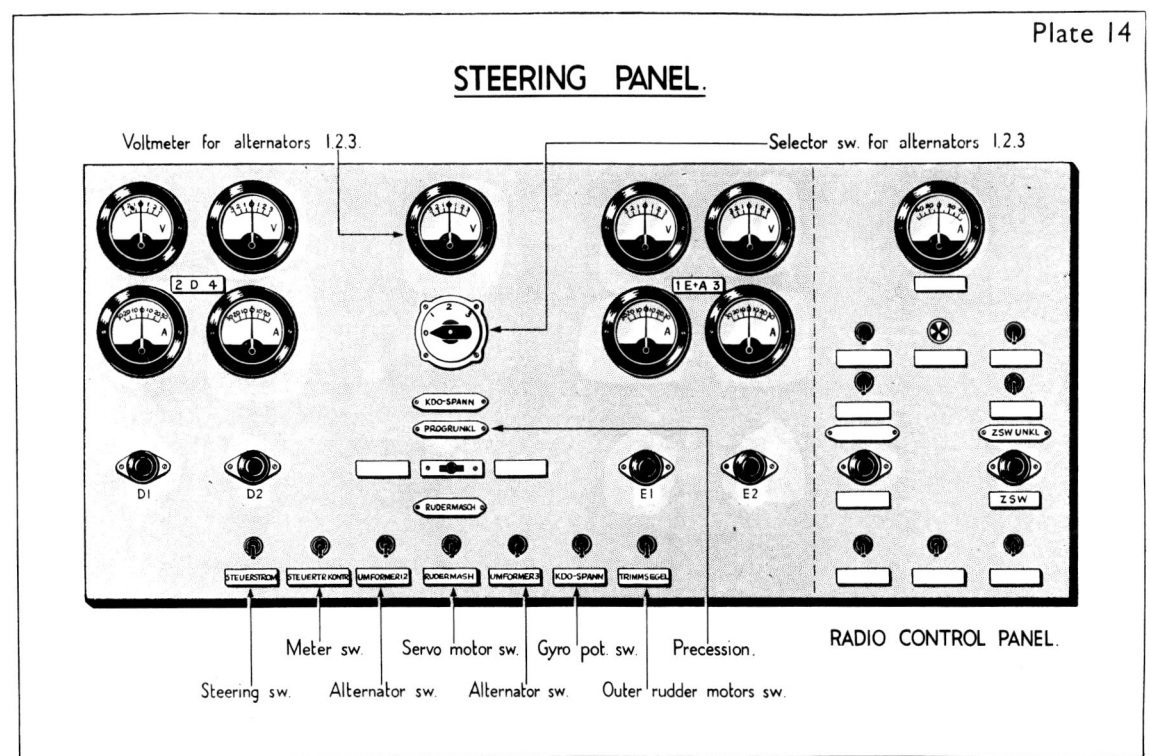

The steering control panel. Launching Troop personnel tested the jet vanes and rudders prior to launch from this panel, and set the electrolytic cell integrating accelerometers. Source: *Report on Operation Backfire*.

The Feuerleitpanzer was a half-track vehicle with an armored cab in back that housed the launch control and missile test panels. Within the vehicle there were panels for steering tests of the missile, rocket motor tests, and launch control. Seats and observation slits with two-inch thick windows were provided for the firing officer and panel operators. There was a communications compartment equipped with a telephone and switchboard. Through the switchboard, communications could be maintained with the Launching Troop Headquarters, the other two firing platoons, and the launch pad. It was generally placed 100 meters from the launch pad, behind the line of fire, and was dug in to the depth of its tracks. Three personnel occupied the Feuerleitpanzer during launch. Other firing platoon personnel were in two-man slit trenches about 150-200 meters from the pad. The ground beneath the launch pad would have been reinforced, frequently with logs.

The Meilerwagen was brought to within 50 feet of the launch pad, at which time the Firing Platoon Truck Section took charge of the missile. After the camouflage cover and rudder protectors were removed, the control compartment batteries, alcohol connection fitting, tools, and other accessories were placed in a box, and hung on a strut at the top level of the lift frame.

Hand winches were used to move the Meilerwagen to the launch pad. Brackets on the pad engaged lugs on the chassis, and indicated the two were properly aligned. After leveling the Meilerwagen by means of two extendable outriggers with end-jacks, the Truck Section began raising the rocket. When the lift frame was vertical, the rocket was suspended just above the pad. Any final adjustments were made to the position of the pad with levers and man power.

The portable launch pad consisted of the launch table, blast deflector, cable mast, valve box, five-way coupling, and oxygen tank topping off connection. The rocket stood on the launch table, which was supported by four tubular legs. It could be rotated, like a turntable, to aim the rocket. To direct the exhaust away from the rocket after ignition, a heavy steel blast deflector was underneath the table. The blast deflector was in the shape of a four-sided concave pyramid. A mast connected to the table carried an electrical cable to the missile control compartment to provide power until launch, when it was cast off. An individual launch pad could be used for 30-40 launches.

During testing, fueling, and launching, the valve box regulated the supply of air pressure to the missile. Electrical connections for the rocket were also contained in the valve box. Terminals for the igniter were on the exterior of the box, on the left side. When traveling, the box was removed from the launch table and carried in another vehicle to prevent damage. It was attached to the table by two brackets. The five-way coupling was an adjustable coupling extending from the valve box to the connection near the base of the rocket between fins II and III.

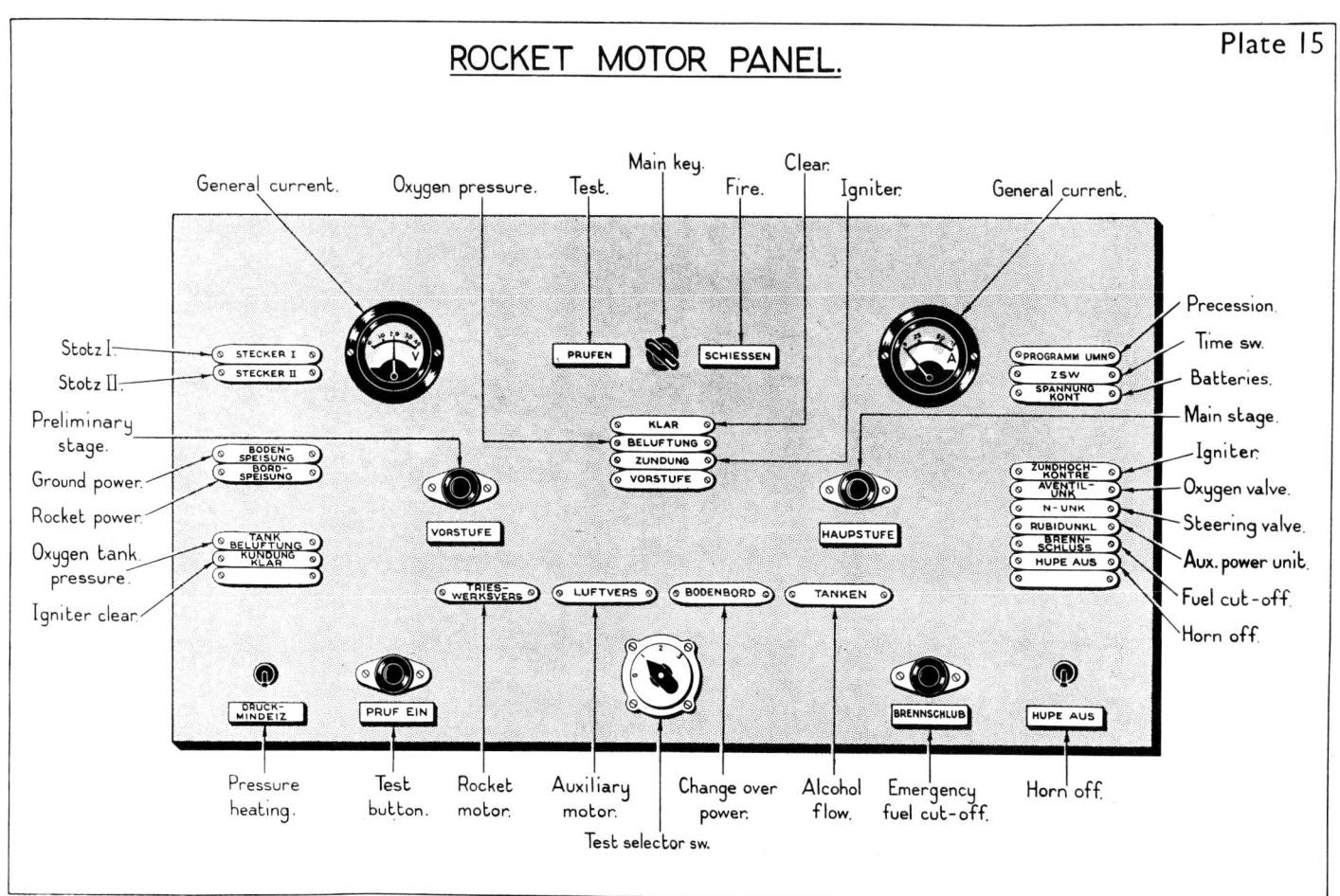

This panel in the Feuerleitpanzer controlled the launch of the V-2. *Source: Report on Operation Backfire.*

Chapter 9: The V-2 at War

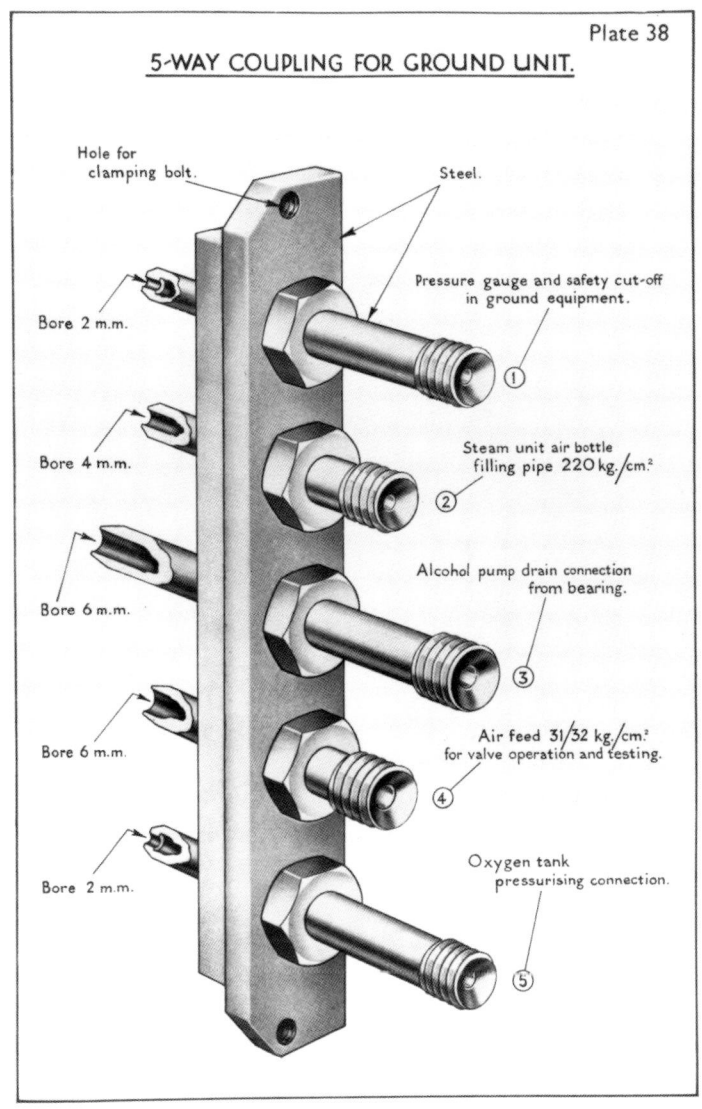

Schematic of the five-way coupling. *Source: Report on Operation Backfire.*

With the missile in the vertical position over the pad, the launch table supporting plates were raised until they just touched the rocket's fins. Next, the clamping collars on the Meilerwagen lift frame were lowered using a ratchet until the rocket rested on the launch pad. After the clamping collars were released, the Meilerwagen was withdrawn about a yard from the rocket. The missile then had to be traversed 90° on its turntable so the fueling connections faced the working platforms. The platforms were lowered, the cable mast raised, and the valve box with the five-way coupling was attached to the launch pad.

Two collimeters, 90° apart and about 100 meters from the missile, were used to ensure that the rocket was perfectly vertical. Necessary corrections were made using jacks in the launching table.

Three of the jet vanes were installed; the fourth was left off until one of the soldiers visually inspected the combustion chamber interior. A wooden platform placed on the blast deflector permitted access to the chamber. Upon completion of the inspection, which was to make sure the propellant injection nozzles were clear of grease and dirt, the platform was removed, and the fourth jet vane installed. Meanwhile, the batteries were installed in the control compartment, and the pressure cylinders in the compartment charged. The five-way coupling was connected to the valve box and the rocket.

Another round of preflight tests followed. Tests were made of the steering mechanism and controls, rocket motor, switch over from ground to internal power, and the rocket's internal sequencing mechanism. These tests were controlled from the Feuerleitpanzer.

A three-section extendable ladder, the Magirus Ladder, was used to reach parts of the rocket inaccessible from the Meilerwagen work platforms. The ladder was mounted on a two-wheel carriage, and could be moved by hand over short distances within the launch area. For movement over long distances it had to be carried on a truck. Fully extended, the ladder reached 56 feet.

The V-2 consumed four propellants: liquid oxygen (A-Stoff); alcohol (B-Stoff); hydrogen peroxide (T-Stoff); and sodium per-

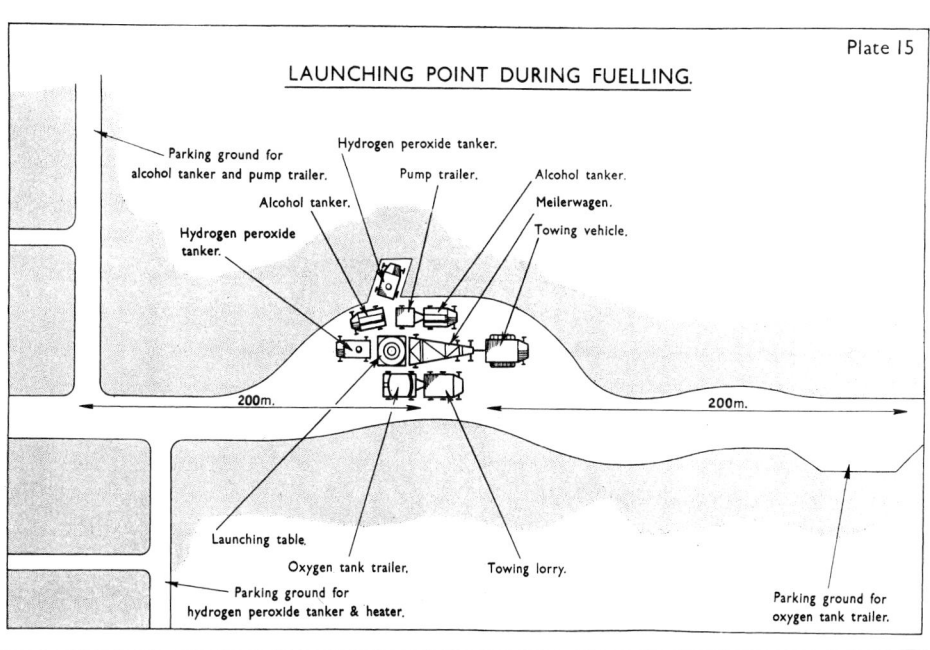

Layout of the firing point during fueling. *Source: Report on Operation Backfire.*

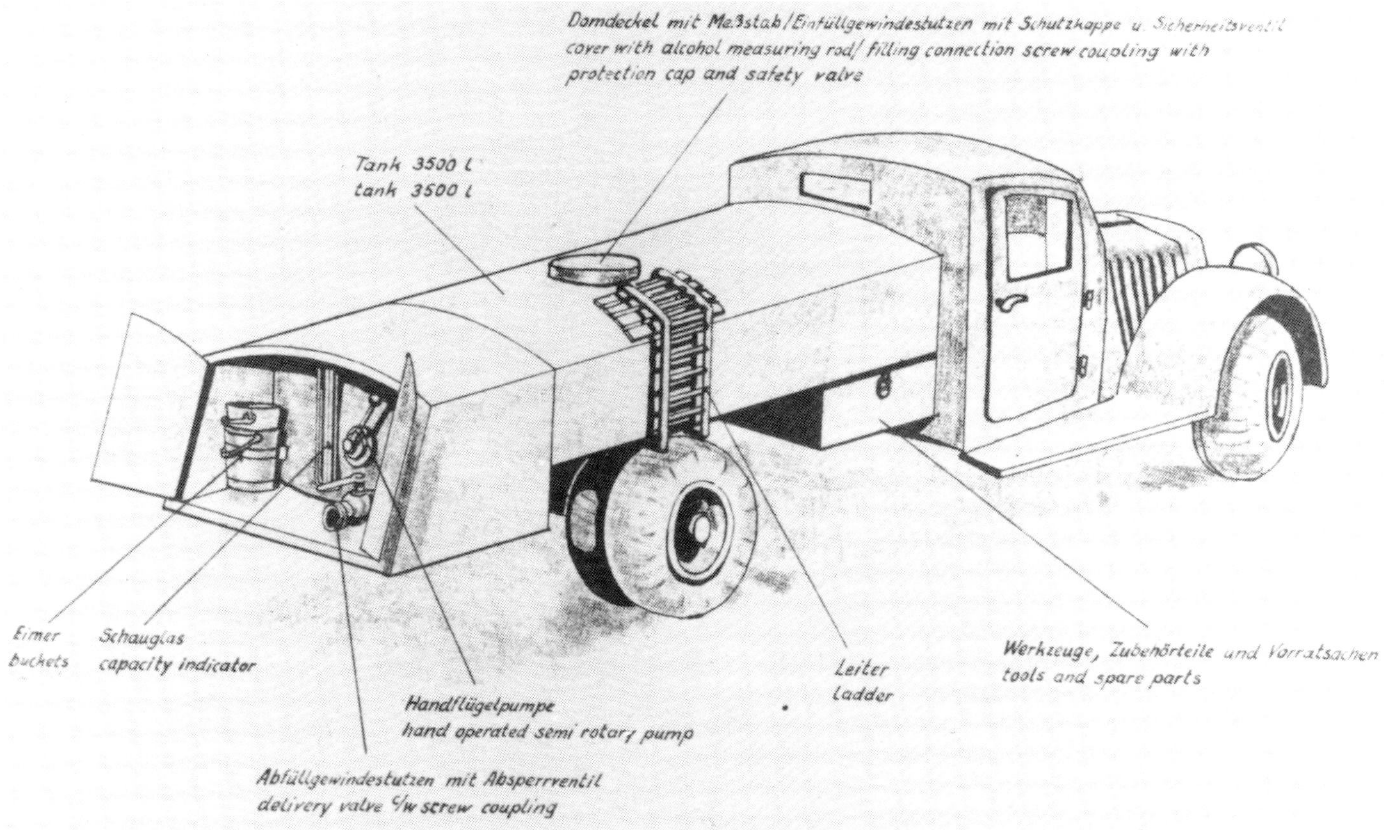

Alcohol, or B-Stoff, tanker. Each rocket required two tankers. *Source: Die Fernrakete.*

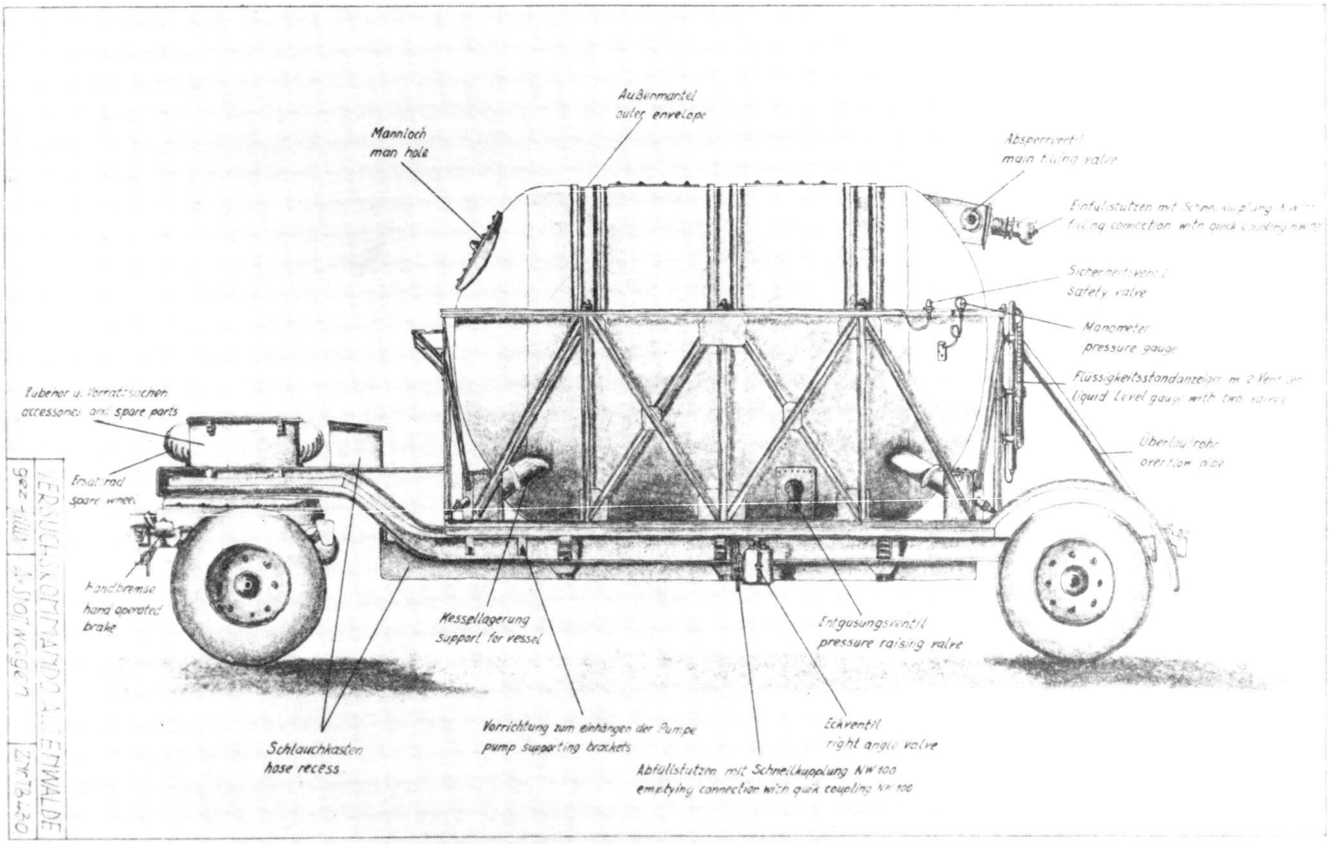

The liquid oxygen, or A-Stoff, trailer. *Source: Die Fernrakete.*

Chapter 9: The V-2 at War

manganate (Z-Stoff). Alcohol and Liquid oxygen were the main propellants; hydrogen peroxide and sodium permanganate were used to generate high-temperature steam to power the missile turbopump. A-Stoff, B-Stoff, and T-Stoff had their own transport vehicles, while Z-Stoff was shipped in containers with the rockets on the special trains.

When the rocket motor and steering tests began, the alcohol, liquid oxygen, and hydrogen peroxide tankers were brought into position, and preparations began for fueling. The seven pressure cylinders in the propulsion unit were also charged. Propellant transport and loading were the responsibilities of the Fuel and Rocket Troop. In the field, they were situated for accessibility to the railway unloading station and main supply routes. One of the main concerns was to avoid congestion in the area as the propellants and rockets arrived. The preferred arrangement was for each battalion to have separate unloading stations for each major supply item, but this was not always possible. The emphasis in deployment of the Fuel and Rocket Troop was to disperse the required vehicles in parking areas along the main supply route.

Alcohol was delivered in tank trucks with a capacity of 2,900 liters (756 gallons). Each rocket launched required two tankers. The fuel tanker comprised an elliptical tank mounted on a 3-ton truck chassis. Transferring fuel from the tankers to the rocket required an external rotary pump powered by a 300-cc single-cycle gasoline engine. One pump could fuel a rocket from two alcohol tankers simultaneously.

After the rocket was fueled the oxygen tank was filled. Liquid oxygen was delivered to the launch site in an insulated container mounted on a trailer. The container was made from an inner aluminum alloy tank with an outer body. There was a layer of glass wool insulation between the alloy tank and outer body. There were only a few plants capable of producing liquid oxygen in large enough quantities to support the missile operation; the highly perishable oxidizer was transported from the plants to the front aboard trains. At the railhead, the liquid oxygen was transferred from the rail cars to the trailers using a portable pump. The same pump was used to fill the rocket tank at the launch site. It normally took about 20 minutes to fill the trailer. Each trailer held 6,750 kilograms of liquid oxygen which, allowing for evaporation, was enough for one rocket. An eight-ton tractor hauled the liquid oxygen trailer.

Five minutes after they began filling the liquid oxygen tank, the Fuel and Rocket Troop began to fill the hydrogen peroxide measuring tank on the Meilerwagen. This tank held 126 liters. Hydrogen peroxide is extremely corrosive, and reacts with organic materials with explosive results, so it must be handled with extreme caution. The transport tanker, which looked like a smaller version of the alcohol truck, had a capacity of 2,120 liters (560 gallons). This was enough for 16 rockets. When the measuring tank on the Meilerwagen was full, the connections between it and the road tanker were closed, and all pump lines flushed with water. The T-Stoff tank in the missile propulsion section was filled from the measuring tank on the Meilerwagen. Finally, the sodium perman-

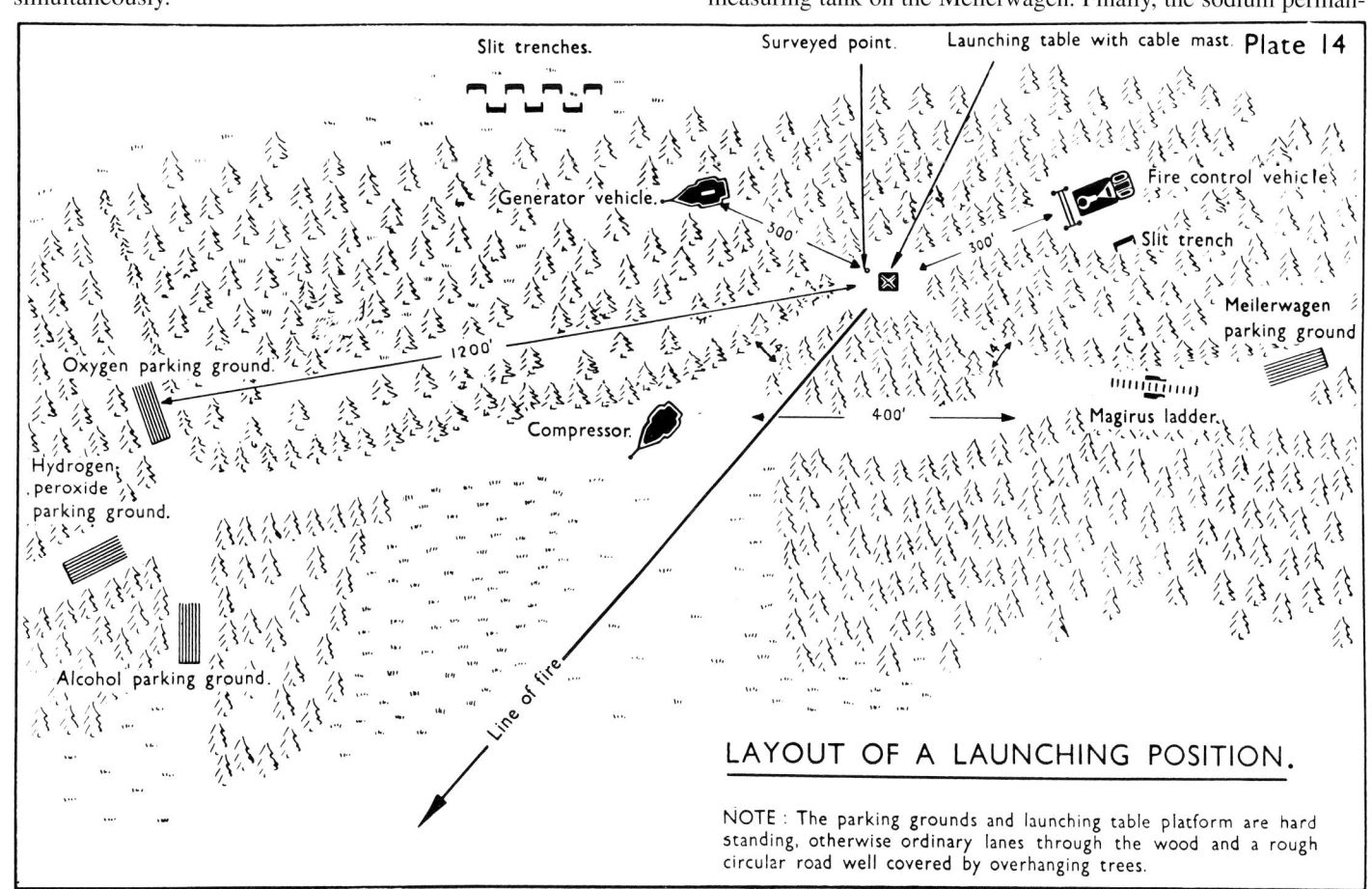

Layout of the firing point during launch. *Source: Report on Operation Backfire.*

Lift-off! This particular photo is of a British post-war test flight of a V-2. *Photograph courtesy Mitchell R. Sharpe.*

Chapter 9: The V-2 at War

ganate, which was in its own container, was poured into the Z-Stoff tank.

Once the rocket was fully fueled, the warhead fuses were armed, all access hatches closed, and final preparations for launching began. The Meilerwagen lift frame was lowered, and the trailer removed from the launch area. If the rocket carried an integrating accelerometer, this had been set for the desired cutoff velocity during the final tests. For radio-controlled cutoff the ground equipment, called Campania, had been set up 6-12 km behind the rocket, in line with the firing direction.

The rocket was traversed so that fin III faced the line of fire, the igniter was installed, and final steering checks were performed from the Feuerleitpanzer. During these tests the gyroscopes were switched on. All vehicles and personnel were evacuated from the immediate area. Once the battery commander was satisfied all was ready, and there were no hostile aircraft in the area, he issued the firing command.

Inside the Feuerleitpanzer, the launch controller turned the firing switch to position two. This closed the relief valves for the A-Stoff, T-Stoff, and Z-Stoff tanks. Compressed air from the ground equipment flowed into the A-Stoff tank to build up pressure between 1.1 and 1.5 atmospheres.

Following tank pressurization, the firing switch was moved to position 3. This opened the control valve from the P-battery of cylinders to force the peroxide and permanganate into the turbopump assembly. Moving the switch to position 3 also switched the rocket from ground electrical power to internal power.

Moving the firing switch to position 4 let the oxygen and alcohol flow into the combustion chamber under their own weight. The igniter fired, and the oxygen and alcohol mixture formed a flickering flame that was called the preliminary stage. During the preliminary stage the motor produced a thrust of about 2.5 tons—far less than what was needed to launch the rocket. This let the fire control officer observe the function of the motor to ensure it would burn evenly.

With the movement of the switch to position 5 the turbopump came up to its full speed of 3,800 RPM, forcing about 72 kilograms of alcohol and 58 kilograms of liquid oxygen into the combustion chamber every second. Motor thrust quickly built up to 25 tons, and the rocket lifted off.

If a fully fueled rocket stood very long on the launch pad, it often created problems due to the super cold liquid oxygen. These problems most frequently consisted of tail section and propulsion unit components freezing. To prevent these problems, Firing Battery personnel had two additional pieces of specialized equipment. The first piece of equipment was a heater for the hydrogen peroxide. Intended for use when the outside temperature was below 20° C, it comprised a double boiler arrangement heated by an alcohol jet. This was seldom used, because nobody liked the idea of using an open flame around concentrated hydrogen peroxide. The second piece of equipment was a hot air blower, which directed a jet of hot air into the tail section of the rocket. This prevented the servo motors and valves from freezing. A two-stroke, two-horsepower engine powered the blower. It was not used in hot weather, or if the rocket was to be launched within an hour.

On 17 March, the 500th SS Battalion fired 11 missiles against the Ludendorf Bridge in Remagen, Germany. One of the missiles hit German held territory 7 miles east of Cologne, and 40 miles short of the target. The remaining rounds averaged a deviation of 1.1 miles in range and 2.5 miles in azimuth. Although this hardly constituted the type of precision attack needed to knock out a bridge, it still is good when one considers that the overall range was 130 miles, and the targeting calculations were hurriedly done in the field.

The last V-2 fired in anger was launched at 8:45 AM on March 28, 1945, towards Antwerp. Himmler ordered the V-2 units disbanded and their personnel reassigned to traditional artillery units. By the time the V-2 campaign ended, more than 1,950 missiles had been fired at targets on the European Continent. Of these, 1,780 were directed at Antwerp. The last London-bound missile had been fired several hours earlier.

On the European continent, the V-2 caused 5,400 deaths, wounded 22,000 people, and destroyed 90,000 homes. An estimated 1,115 V-2s struck England, half (518) of which hit London and its suburbs. There were 2,754 deaths and 6,523 serious injuries in London alone from the barrage. The 2,344 V-1s that struck the city killed 5,143 and injured 14,756.

The V-2 caused a great deal of misery and destruction, but in its mission as a weapon system, it failed to live up to expectations. From the German perspective, this was a very costly weapon to deploy. Each firing battery, which could launch up to nine missiles per day, required over 500 men using 152 self-propelled vehicles, 70 trailers, and 5 bicycles. Nine missiles could deliver 7 1/2 tons of explosives, less than the maximum load of a single B-17 bomber. When compared with the destruction being delivered each day and night by the bombers of the American Army Air Force and British Royal Air Force, the V-2 pales in comparison. While it did have an impact on the Allies—manpower had to be diverted from other necessary offensive military operations to civil defense, aerial reconnaissance, and bombing of flying bomb sites—England was not terrorized into surrender, and the flow of supplies through Antwerp and Liége was barely affected.

However, despite its failure to alter the course of World War II, the V-2 stands out as a major technological achievement. As the world's first large ballistic missile, the V-2 was a harbinger of things to come, and the victorious allies were eager to study it. V-2s would continue fly, not over the English Channel, but over the steppes of Russia and deserts of the southwestern United States.

10

European Round Up

With the collapse of the Third Reich, the Allies scrambled to recover as many of Germany's advanced weapons as they could. After all, the Germans had deployed jet aircraft, long range rockets, and other weapons far ahead of anything previously seen. Even more important were the scientists and engineers who developed these weapons, and the victors set up programs to employ their former adversaries to exploit their knowledge. The British, Americans, and Soviets were especially enthusiastic in their recovery of missile hardware and recruitment of German personnel

In the final weeks of the war in Europe, elements of the advancing Allied armies found V-2 launch sites and equipment throughout Western Europe. Holland was a particularly rich area. Canadian forces operating near Hellendoorn, Holland, found the site previously used by the 500th SS Battalion. Local residents estimated about 160 rockets had been launched from there before the Germans moved out on 20 March. Craters in the surrounding area revealed that some rockets crashed shortly after take off.

British units advancing through Hannenberg, Germany, found an outdoor missile dump. Dr. Georg Rickhey, General Manager of

Canadian forces moving through Hellendoorn, in Holland, discovered log reinforced platforms used by V-2 firing batteries. Craters and missile fragments in the vicinity also showed that not all launches succeeded.

Chapter 10: European Round Up

Mittelwerk, estimated there were as many as 2,100 V-2s in similar storage sites throughout Germany when the last operational firing occurred on March 28, 1945. Slightly over half the missiles were in the Soviet occupation zone, but 515 were developmental models consigned to salvage. Rickhey estimated another 1,000 missiles were scattered throughout the British, American, and French occupation zones.

Disposition of V-2 Rockets as recalled by Georg Rickhey:

Reaching England	1,115
Reaching Continental Europe	1,117
Airbursts	600-700
Total Launched	3,600
In Mittelwerk at time of occupation	250
In field storage in English, American, and French zones	1,000
In field storage in Soviet zone	1,100
Total accounted for	5,950

Source: V-2 Rocket Attacks and Defense.

Nearly all the missiles found in the storage areas were incomplete. Most were missing critical control components, which were necessary to guide them as they flew. Great Britain sought to acquire 150 missiles. Thirty were for their Army's flight research program, while the rest were for evaluation by the Air Ministry. The United States wanted at least 100 rockets. The French and Soviets also sought to recover missiles. Competition became so intense that on 4 May, orders were issued to Allied units freezing all missile hardware in Europe until the question of allocating the treasure trove could be resolved. This order seems to have been largely ignored.

Elements of the United States Army occupied Mittelwerk in April, and found about 250 rockets in various stages of completion on the assembly line. Large quantities of components were also found stacked up outside the factory. No complete rockets were found. Working under the supervision of Major James Hamil, there was a hurried effort to evacuate as much materiel from the underground factory as possible, because Nordhausen was scheduled to be placed under Soviet control. In general, this was limited to rockets and components; production machinery was left behind. Ulti-

The British Army found several storage areas and depots for missiles like this one near Leese. There were many rockets in these depots, but most had been destroyed or damaged by retreating German forces.

An American Military Policeman inspects a partially completed rocket on the assembly line at Mittelwerk. The Americans occupied the factory, and shipped an estimated 640 tons of materiel back to the United States. *Source: U.S. Army photograph.*

Inspecting a V-2 on a rail car on its way to the front. *Source: U.S. Army photograph.*

Chapter 10: European Round Up

mately 640 tons of equipment, requiring 300 railroad cars, was taken to Antwerp, and loaded upon Liberty Ships for shipment to the United States. After an official British protest, six tail units needed for their program were off loaded before the ships departed the Belgian port. The British needed the tail units to complete their missiles.

Peenemünde would have seemed to be the top prize, but when the Russian Army captured it in March the facility resembled a ghost town, and all the senior personnel were gone. Fearing capture by the Russians, von Braun and his top people had long since left. They first moved to Bleicherode, near Nordhausen, then finally migrated south to a mountainous area near the German-Austrian border. This second move was not of their own choosing—SS General Kammler had ordered them to report to Oberammergau.

When the rocketeers fled Peenemünde they took their technical records with them. These records contained the results of nearly 15 years of experimental work. Whoever recovered the archives could begin working where the Germans left off. Realizing the incalculable value of the documents, and believing they might be useful as a bargaining tool later, von Braun detailed Dieter Huzel and Bernhard Tessman to hide them.

Huzel and Tessman found an unused mine near Goslar, in the hills north of Nordhausen. Since the precise location was to be a secret known only to the pair, the soldiers who helped move the documents stayed in the backs of the trucks as they traveled to the site. As far as the owner of the mine was concerned, he was told that they were classified documents that needed to be safeguarded. After the documents were placed in one of the mine's galleries, the shaft leading to it was sealed using dynamite.

Their job accomplished, Tessman and Huzel split up. Tessman rejoined von Braun and the other Peenemünde scientists at Bleicherode; Huzel remained in the Goslar area to check the results of the blasting job. He remained for two days, visiting an old friend, then left for Bleicherode upon hearing American troops were approaching.

By the time Huzel reached Bleicherode, von Braun and the others had begun their flight south. Huzel detoured to Berlin to remove his fiancée from the path of the rapacious Red Army, then headed to rendezvous with von Braun. After traveling several days he reached his comrades at Oberammergau, near the Austrian border. Despite the idyllic alpine setting, Huzel discovered that all was not well in Oberammergau.

Kammler had posted SS guards on the rocket group. In addition to his duties as titular head of the rocket program, Kammler had been appointed to head turbojet fighter production. He had summoned the Peenemünde personnel, which numbered about 400, and a group from Messerschmitt to the alpine town. Fearing the SS, the rocket group opted to migrate once more, to the Haus Ingeborg, in Obejoch, near the Adolf Hitler Pass. To do this they had to escape from their SS guards.

They eluded their guards largely because Kammler was no longer on the scene. In his memoir *V-2*, Dornberger described how he last saw Kammler in early April when the evacuation from Oberammergau began. In the last weeks of the war Kammler refused to admit defeat. With fanatic zeal he drove himself and those around him, always professing his belief that Germany could still win the war. Meetings were frequently held late at night and along highways. He was directed to command the defense of Czechoslovakia, and there he disappeared. One of the most plausible theories as to his fate is that his adjutant shot him to prevent his capture.

After two weeks at the Haus Ingeborg, spent mostly telling stories and playing chess, the group decided to surrender to the

Wernher von Braun and his team of rocket scientists surrendered to the American Army near Ruette, Austria. Left to right: Charles L. Stewart, U.S. Counterintelligence Corps; Lt. Col. Herbert Arter, a member of General Dornberger's staff; Dieter K. Huzel (in black hat); Wernher von Braun; and Magnus von Braun. Wernher von Braun's arm is in a cast from a car accident several months earlier. *Source: New Mexico Museum of Space History.*

During the summer of 1945, the British Army assembled V-2 missiles and ground equipment at the Krupp Proving Ground, near Cuxhaven. The British code named the effort Operation Backfire. *Source: Mitchell R. Sharpe.*

Americans. Wernher von Braun's brother Magnus was dispatched on a bicycle to find American forces rumored to be operating near Reutte, Austria. Magnus spoke the best english of the group, and they decided he could best negotiate their surrender with the American Army. On May 2, 1945, they surrendered to elements of the 44th Infantry Division. Several days after their surrender the Germans were taken to Garmisch-Partenkirchen for interrogation and processing. During the interviews the location of the hidden documents was revealed, and these were recovered and sent back to the United States.

Early in May the British War Ministry decided to conduct test firings of V-2s using German personnel. During the summer and fall of 1945 the British mounted a major program to investigate the V-2. Called Operation Backfire, the effort sought to study V-2 field operations. Backfire was headquartered at the Krupp Naval Gun Testing Ground on the North Sea coast, near Cuxhaven. Unlike the planned American and Soviet V-2 programs, Backfire sought to duplicate (as closely as possible) wartime firings so German field procedures could be studied.

The Supreme Headquarters Allied Expeditionary Force tasked the British 21st Army Group with providing a brigade for Backfire. The 307th Infantry Brigade, commanded by Brigadier L.K. Lockhart, received the assignment. Upon dissolution of the Supreme Headquarters Allied Expeditionary Force on 14 July, the headquarters responsible for Backfire became the Special Projectile Operations Group (SPOG). Sir Alwyn Crow, who participated in the Big Ben investigation of the V-2, oversaw Backfire for the British Ministry of Supply.

Collecting the necessary hardware proved a more difficult task than first anticipated. The British presumed they would find complete, ready to fire missiles, but this was not the case. None of the rockets they found were complete. Most had been stored outdoors, and there was considerable deterioration, particularly in the electrical components. They also discovered that retreating German soldiers tried to render the equipment unusable. Local civilians who, once liberated, were eager to lash out at the former Nazi war machine also took their toll on the rocket hardware.

Unable to find any complete rockets, the SPOG devoted more than 200,000 man hours with a work force of over 200 people in the search for V-2 parts. Their search covered half a million miles, and required 100 vehicles and 400 railway cars! The SPOG employed 600 Germans. Only 128 former military personnel and 79 civilian technicians had any prior experience with the V-2. The rest were troops and civilians with no previous involvement with the missile.

Once the parts were secured, the British encountered further difficulties assembling the missiles, for the Germans would not improvise or make do with what was available. Instead, they frequently demanded special tools, which complicated matters.

The British Army originally hoped to launch about 30 rockets to test various control methods. They also intended to provide 120 missiles to the Air Ministry for study, but due to shortages of critical parts (particularly gyroscopes, batteries, and other control components) only 8 complete missiles could be assembled. Of these only three flew. The military and civilian contingents were segregated so their answers to technical questions could be checked against one another. The soldiers and civilians who prepared and launched the rockets comprised the *Altenwalde Versuchs Kommando* (AVKO). The remaining civilian experts were housed at Brockeswilde. AVKO was organized as a military-style unit. Lieutenant Colonel Wolfgang Weber, former commander of one of the V-2 regiments, was its head.

At first, General Dornberger was present at Cuxhaven, but was kept away from both groups. After a few weeks he was taken to England to face charges as a war criminal. Dornberger was charged with launching V-2s against the open city of London. In his defense, it was argued that he was only in charge of the missile's development; Kammler directed their use. After two years

Chapter 10: European Round Up

Dornberger was acquitted, and offered a contract to work for the United States Air Force. Still later, in May 1950, he went to work for Bell Aircraft.

Even the field equipment recovered by the British was in poor condition, because it had been stored outdoors for several months without any maintenance. Refurbishing the Meilerwagens, Vidalwagens, launch pads, propellant trucks, and other support equipment became nearly as complex as preparing the rockets themselves.

Six British technical officers supervised the Germans who assembled and tested the rockets. There were also two American observers present. By mid-fall, the SPOG had the first missile ready. On October 1 the SPOG attempted to launch the first rocket. After two unsuccessful attempts, the missile was removed and taken back to the shop area for further work.

The next day the second rocket flew successfully. It impacted in the North Sea, less than a mile and a half from the intended target. On 3 October the rocket from the first launch attempt finally flew. Something went wrong, however, and the fuel cutoff occurred only 34 1/2 seconds after launch. This rocket only traveled 15 miles.

The third Backfire launch took place on October 15. Code named Operation Clitterhouse, this flight was a demonstration for representatives from the United States, France, Soviet Union, and press. Despite miserable weather conditions the rocket performed well, impacting within 12 miles of the target.

A week after the Clitterhouse launch, the SPOG began processing the German civilians who worked on Backfire for release. Some had already been offered contracts in the United States. About 20 accepted offers to work in Great Britain; the rest returned to their homes in Germany. The British were quite thorough in their

Raising one of the Operation Backfire missiles. The British launched three V-2s over the North Sea in October 1945. Gathering and refurbishing the ground support equipment proved as difficult as assembling the rockets. *Source: Mitchell R. Sharpe.*

German technicians apply a paper drawing to the base of the first V-2 fired during Operation Backfire. The tradition of placing illustrations on test missiles began at Peenemünde in 1942. The first Backfire launch occurred on October 2, 1945. *Source: Mitchell R. Sharpe.*

reporting. The original typewritten report, a complete manual for preparing and launching the V-2, comprised over 300 pages, included hundreds of illustrations, and weighed 20 pounds! In published form, the Operation Backfire report comprised five volumes. Interestingly, during the war the Germans never prepared such a manual.

The British did not launch any additional V-2s, but several members of the British Interplanetary Society proposed adapting the German missiles to human space flight. Founded in 1933 by a group of space enthusiasts, the British Interplanetary Society (BIS) remains, to this day, a lively forum for those interested in space exploration. At the Society's January 7, 1948, meeting, BIS fellows R.A. Smith and H.E. Ross presented a proposal for a suborbital "man-carrying rocket." Smith and Ross submitted this project to the British Ministry of Supply in late 1946, but it was rejected.

They suggested building a rocket based on the V-2, topped with a manned cabin instead of a warhead. Such a rocket would require an engine only slightly larger than the one used in the V-2, and Ross and Smith proposed:

"...that a rocket should be constructed utilising a motor similar to that employed in the A.4 series of German turbo-rockets."

The V-2's engine burned alcohol and liquid oxygen to produce a thrust of 54,000 pounds. The man-carrying rocket required an engine capable of producing 60,000 pounds.

To save weight the fins were omitted; instead, the rocket relied on graphite vanes in the motor's exhaust for steering. The rocket was also slightly longer and fatter than the V-2. It would have been capable of reaching 200 miles, but Smith and Ross recommended:

"...that the full period of thrust should not be utilised because of the danger of destruction of the cabin by the powerful deceleration resulting from a high-speed drop through the barometric gradient."

During launch, the cabin was enclosed in a fairing. After powered ascent the fairing peeled away to expose the capsule. Centrifugal force would aid jettisoning the fairing pieces. When the rocket reached the desired velocity, the propellant valves were to be closed, shutting off the engine, but the propellant turbopump would be allowed to continue running. Torque from the turbine would make the rocket spin about its long axis. Once the spinning reached a high enough rate, the fairing pieces were cast off. The pilot would allow the spinning to continue until centrifugal force produced a 1-g acceleration on his body. At that point he was to shut off the propellant turbine and let the spin subside due to friction in the turbine bearings.

His next act would have been to release a compressed air charge to separate the capsule from the booster.

"The pressure cabin would be constructed of a suitable light metal alloy sheeting, and would have two ports with self-sealing conical gaskets for access."

Four hydrogen peroxide thrusters could control capsule orientation. The recovery parachute was packed in a container mounted above the capsule on tubular supports. It was intended to deploy the parachute early in the descent. The parachute needed to be a "constant drag" type that provided the same retarding force regardless of air density or speed.

Ross and Smith even suggested a development program, beginning with propulsion tests and ending with the piloted flight. One of the interesting features of their program was the "Operational drill," where the pilot would:

"...be called upon to guide a telecontrolled rocket by instruments similar to those he will use in the flight, including controls for nose-fairing and parachute release."

During the flight, the pilot was to conduct observations of the Earth and Sun, test radio communications through the ionosphere,

Chapter 10: European Round Up

and gather data on human response to acceleration and weightlessness.

The Man-Carrying Rocket was not the only proposal by Ross and Smith. They designed lunar spacecraft, a space station, and a space suit for use on the Moon. However, like the Man-Carrying Rocket, these remained paper studies.

An interesting scene played out just before the Clitterhouse launch, when the Soviets tried to bring in two more people than were on the pre-approved list of observers. The British remained adamant, and the two were denied admission to the Krupp grounds, so they could only watch the launch from outside the fenced perimeter. One of the Soviets was an engineer named Sergei Korolev, who wore the uniform of a Captain, although his actual rank was Colonel.

Unknown to either the British or Americans, Korolev was the head of the Soviet long range rocket program. The leader of the Soviet effort was Sergei Pavlovich Korolev. Born on December 30, 1906, (Julian, or old-style calendar; using the Gregorian calendar his birth date was January 12, 1907) in Zhitomir, Ukraine, Korolev was fascinated with airplanes as a boy. He grew up to be an aircraft designer at the Tupolev Design Bureau, and was particularly enthusiastic about gliders. In the early-1930s Korolev joined the Mos-

Final preparations of one of the Operation Backfire V-2s. Note the mast erected alongside the missile. This carried electrical cables to the control compartment that disconnected as the rocket lifted off. *Source: Mitchell R. Sharpe.*

Another view of an Operation Backfire V-2 on the launch pad. This view shows the hydraulic cylinders and lift frame of the Meilerwagen. This photograph is of the first attempted launch on October 1, 1945. After two unsuccessful attempts that day, the rocket was removed from the pad and taken back to the workshop. *Source: Mitchell R. Sharpe.*

Chapter 10: European Round Up

cow-based Group for the Study of Reactive Motion, or GIRD, primarily to develop propulsion for his gliders.

Fredrikh Tsander founded the Moscow organization in late 1931. A similarly titled group also started in Leningrad at about the same time. To distinguish the groups, they became known as MosGIRD and LenGIRD, respectively. Soviet Armaments Minister Mikhail N. Tukhachevskiy noted the work of the GIRD organizations and supported them. With official support, MosGIRD hired 10 full time employees, including Korolev. By May 1932 Korolev headed MosGIRD. He and Tsander had several projects in progress that led to the first Soviet liquid fuel rockets. Tsander's untimely death in 1933 from typhus was a major loss to the fledgling organization.

Under Korolev's leadership, MosGIRD built and launched the GIRD 09 on August 17, 1933, from Podmoskovia, near Moscow. GIRD 09 is best described as a "hybrid" rocket, since it burned a mixture of gelled gasoline and liquid oxygen. Mikhail Tikhonravov headed the technical team within MosGIRD that designed and built the rocket. Korolev used an engine developed by Tsander before his death to propel the GIRD-X on November 25, 1933. Burning liquid oxygen and ethyl alcohol, this was the first all-liquid fuel rocket in the Soviet Union.

Other groups of rocket and space travel enthusiasts popped up in the Soviet Union during the 1920s and '30s. The most significant was the Gas Dynamics Laboratory (GDL) in Leningrad. The immediate predecessor of the GDL began operating in 1922 in both Moscow and Leningrad under the name "Laboratory for Development of Engineer Tikhomirov's Invention."

Tikhomirov was a chemical engineer who began experimenting with solid fuel rockets in 1894. He submitted his ideas to a military panel in 1916, but it was not until after the Communist Revolution and Civil War that he received any funding. Given a modest machine shop and a staff of 10, his initial work involved smokeless powder artillery and airplane takeoff rockets.

The laboratory consolidated its operations in Leningrad around 1925. In June 1928 it became the Gas Dynamics Laboratory (GDL). The following year, 21-year old Valentin Glushko, who attended Leningrad University before he joined the GDL, suggested they add liquid and electrical rocket subdivisions. Glushko became head of the effort. As a teenager Glushko read the works of Jules Verne, which inspired a life-long interest in space travel. He wrote to Tsiolkovsky, and told him of his interests in astronomy and astronautics. Tsiolkovsky encouraged him to pursue his dream. (Glushko was a member of the Soviet delegation admitted to the Clitterhouse launch.)

In 1933 the GDL merged with MosGIRD to form the Jet Propulsion Scientific Research Institute (RNII), an entity supported by the Armaments Minister. A highly successful result of the creation of the RNII was the use of Glushko's ORM-65 engine in the RP-318 glider designed by Korolev.

The group continued to progress until the spring of 1938, when Tukhachevskiy was arrested and executed during Stalin's purges. Those who worked for the former minister were also arrested, including Glushko and Korolev. Glushko, who was arrested in March, became one of Korolev's accusers. One night in June the dreaded knock on the door occurred, and Korolev was taken into custody. Korolev was initially sentenced to 10 years at hard labor in Siberia, followed by five years of disenfranchisement and forfeiture of all personal property. He ended up in the Kolyma Gold Mines, which were particularly harsh, even by gulag standards.

Korolev survived in the camp for about a year, when he was summoned and sent to a special prison near Moscow. Realizing that many of the inmates in the gulags had valuable skills, Stalin authorized special prisons called "sharashkas," where their talents could be used. Tupolev, who was also under arrest, requested that Korolev be transferred to one of these as an aviation scientist. Glushko had been similarly imprisoned in a sharashka and allowed to work on rocket propulsion. During the war, as German troops advanced, Tupolev's sharashka relocated to Omsk.

Churchill informed Stalin about the V-2 as early as July 1944, and asked the Soviets if they could help get a group from England into the area around Blizna to investigate. Stalin dispatched a team of Soviet specialists to gather and examine rocket hardware. They recovered rocket components, and shipped them back to Moscow for study. The Russians were amazed at the German advances in rocketry.

Recognizing the likelihood of rockets being an important weapon in future conflicts, the Soviets launched a full-scale effort to capture as much materiel as they could from the German rocket program. Soviet ground forces captured Peenemünde on May 3, 1945. As previously described, they found the installation occupied by middle level managers and technicians, since von Braun and most of his senior staff had already fled west and surrendered to the Americans.

The most senior manager from Peenemünde who went to the Soviets was Helmut Gröttrup, who was deputy to Dr. Ernst Steinhoff in the guidance and instruments laboratory. Gröttrup was not satisfied with the terms of the contract offered by the Americans under Project Paperclip. Accepting the Paperclip contract would have required him to move to the United States without his family. He also was not pleased by the fact that the Army could terminate it at any time. Gröttrup and his wife decided to stay in Germany, and he settled in Bleicherode. Chertok, who headed the Russian group at the time, sought Gröttrup, and offered him terms that (at the time) seemed more to his liking if he worked for the Soviets. Kurt Magnus, a specialist in gyroscopes, also joined the Russians.

The Americans managed to come away with all the major prizes—the factory, senior personnel, and technical archives. The Mittelwerk complex was in an area slated to be turned over to the Soviets, so the American Army removed more than 300 boxcar loads of hardware from the factory. By the time the facility was turned over on 5 July, only components and machine tools remained. When told of how the Americans managed to secure the top rocket scientists, technical archives, and had first crack at the factory, Stalin was furious.

The first Soviet rocket personnel arrived in Germany just behind the front line troops. They set up headquarters in Bleicherode in late April. Korolev, who had been released from prison in 1944, arrived at Mittelwerk in September 1945 to restart the production line. Although they only had components of missiles and little technical data to draw upon, the Soviets decided to replicate the V-2.

Korolev set up the Nordhausen Institute at Bleicherode, and began the task as its Chief Designer. Although he was still technically under arrest, Korolev was made a Colonel in the Red Army before being dispatched to Germany. Besides the V-2 factory, Soviet forces captured the German liquid oxygen plant and put it back into operation. Glushko was also sent to Germany to construct a test stand near Lehesten. By early September Glushko made his first static firing of a V-2 engine. Glushko then copied and improved the V-2 engine. His improved V-2 engine bore the Soviet designation RD-100.

On May 13, 1946, the Soviet Council of Ministers and Party Central Committee issued a formal decree that governed the way rocketry would be pursed. The decree comprised three provisions. The first designated the Ministry of Armament as the responsible agency for the development and manufacture of missiles. Dimitri Ustinov was appointed minister. The second decree formalized the work that had already been done with the Nordhausen Institute. Major General L. M. Gaidukov was appointed to head the Institute, with Korolev as his lead engineer. The third step called for the establishment of a testing range for the launching of large missiles.

Soviet technical teams found blueprints and missile specifications in a German archive in Prague. Korolev asked Vasily P. Mishin to head the team that translated the German documentation into Russian. For two years Korolev labored at the Nordhausen Institute, carefully piecing together V-2s. Korolev had a flyable rocket by 1947, so he closed the Nordhausen Institute and moved the production line to the Soviet Union.

There were about 1,000 people working on the nascent Soviet rocket program in Nordhausen, and about half of them were Germans. Korolev did not start his assembly line in Mittelwerk. Instead, he preferred an above ground site, and used a V-2 repair depot called Klein Bodungen, which was nearby. When Korolev closed the Nordhausen Institute all the Germans were forcibly moved to the Soviet Union. They were well paid, however, and were allowed to take all their household belongings with them. Fearing food shortages, the Grötrupps insisted on taking their cow and a supply of hay with them. The Soviets acquiesced to this demand, and loaded the cow and its fodder on the train heading east.

Korolev rejoined Glushko, Tikhonravov, and many of his earlier associates at a newly designated design bureau, NII-88. Scientific Research Institute No. 88 (NII-88) had been created in August 1946, in Podlipki, which became known as Kaliningrad. Three distinct organizations existed within NII-88. The first was an experimental plant; the second was a Special Design Bureau, or SKB; and the third was a group of laboratories that worked on such areas as aerodynamics, propulsion, guidance, and telemetry. Within the SKB, Korolev headed the division charged with developing ballistic missiles, SKB Number 3. In 1950 SKB Number 3 became Experimental Design Bureau (OKB)-1. Six years later, OKB-1 became independent of the NII-88 organization.

Soviet military officials established a State Central Testing Range at Kapustin Yar for missile testing. Located about 75 miles south of Volgograd, in the pre-Volga Steppes, conditions were quite primitive at first. When the first construction crews arrived, there was literally nothing there but a vast, empty wasteland. Temperatures ranged from 40°F below zero in the winter to 100°F above in the summer. During warm weather spiders and poisonous snakes were constant hazards, and clouds of gnats tormented people. Dust blew everywhere. Because there were no permanent facilities, Korolev and his people worked out of specially outfitted trains that had a kitchen, restaurant, bath, library, dispensary, storage facilities, living quarters, laboratories, workshops, and a club with a piano. These trains came from Germany, and had to be refitted with different gauge wheels when they crossed into Soviet territory.

Korolev and the scientists lived aboard the trains. The ordinary soldiers assigned to build the base were not so fortunate. They lived in tents and dugouts. Despite the hardships morale remained high. Part of this was no doubt due to the belief that what they were doing was a matter of utmost importance to their country, but an even larger measure of credit must be given to the leadership style of Korolev. He was a stocky, powerfully built man with dark eyes that burned with intensity. Not one to suffer fools, Korolev could be harsh at times, and was capable of towering rages and blistering rebukes. Yet, at the same time, he was quite caring for his subordinates, and gave them wide latitude to express their opinions when they differed with his. When wrong, he was quick to apologize—a trait that helped inspire a fierce loyalty among his workers. A talented and insightful engineer with an intuitive ability to solve complex technical problems, Korolev was also a skilled administrator. He applied a systems engineering approach to missile development, managing the myriad of steps and processes that had to come together in the end.

On April 14, 1947, a conference was held at the Kremlin to establish the direction of future Soviet missile programs. It was decided to concentrate on two or three designs. The Soviet produced V-2s were designated R-1. An improved version with nearly twice the range was called the R-2. Korolev and the NII-88 team launched the first Soviet assembled A-4 from Kapustin Yar on October 18, 1947. Although this was designated an R-1, it was a V-2 assembled from German parts. The first 11 R-1s were built like this; they did not launch the first all-Russian manufactured R-1 until September 7, 1948. With the Soviet manufactured missiles Korolev made some minor modifications to the design.

The first two missiles veered off course, much to the consternation of the Russians. The Germans determined the problem was caused by vibrations affecting the gyroscopes, and was quickly fixed. R-1 number 3 flew perfectly, and impacted close to the aiming point. At first the Soviets rewarded the Germans with a cash payment and bottle of alcohol. Celebration quickly turned to suspicion, though, as the Soviets began to question why the problem was so quickly solved. Soviet Armaments Minister Ustinov ordered Korolev to investigate whether the failures of the first two missiles was due to sabotage. Korolev quickly learned such failures were within the norm.

Grötrupp was appalled over the Soviets' apparent lack of concern for safety. He witnessed several fatal accidents, and was shocked when there was no reaction among the other Russian work-

Chapter 10: European Round Up

ers. When the first R-1 was nearing launch one of the legs of the launch platform collapsed, and the fully fueled missile began to topple over. Grötrupp ran for cover, and was amazed to see the Soviet workmen running *towards* the rocket. He watched with further amazement as the Russians winched the rocket and platform back into position, and quickly braced it with girders. The launch then proceeded.

During the first firings the Germans worked alongside the Soviet team, advising them. Gradually, though, the Germans were phased out, and eventually they were isolated and worked on paper studies only. By June 1948 the Soviets had moved most of the German rocket specialists to Gorodomlya Island, in Lake Seliger, which is north of Moscow. The Soviets asked them to work on a number of rocket projects that paralleled those being developed by Korolev, but none of these ever progressed beyond the design stage. Gorodomlya was a difficult place to be for the Germans. Some of the bitterest fighting on the Russian Front had taken place in the area, and the local residents were openly hostile towards the Germans. The fact that the German rocket specialists were better paid than the locals didn't help matters, either.

Korolev's next task was to improve upon the R-1. By lengthening the propellant tanks, improving the propellant turbopumps, adding a detachable warhead, and other changes to the R-1, Korolev devised the R-2, with twice the range of the first missile. However, as tactical missiles suitable for widespread deployment the R-1 and R-2 had many shortcomings. Their warhead carrying capacity was too small, their range was too short, and they burned liquid oxygen, which was extremely perishable, especially in a field situation. As it turned out, they made better carriers for instruments than explosives.

Korolev modified the R-1 into a "geophysical" rocket by equipping it with a detachable warhead and external containers for instruments. Designated the R-1A, Korolev first flew this modified rocket on April 21, 1949. Five months later he had the first R-2 ready for flight. By adding a parachute Korolev easily adapted the detachable R-2 warhead into a recoverable payload section, and used it to carry live payloads. In 1951 he flew two dogs aboard an R-2, and recovered them after they reached more than 60 miles. The dogs were named Dezik and Tsygen. After the flight Tsygen became the pet of Anatoli A. Blagonaravov, who became Chairman of the USSR Academy of Sciences Commission for Space Research and Utilization.

Korolev selected dogs because they were readily accessible, and would survive the harsh Russian winters better than more tropical animals like monkeys, which would become the test animal of choice for the Americans. Dogs also had an advantage of being reasonably small in size. Soviet scientists preferred mongrels rather than pure-breeds, because they were more even-tempered. Waste management proved easier with females, so none of the dogs launched by Korolev were males.

For the initial series of flights nine dogs were trained for the rigors of rocket flight. Their training included having them sit for increasing periods of time in a simulated capsule. Animals that did not acclimate to confinement within the capsule were removed from the research program. Those that remained were then exposed to some of the anticipated environmental conditions, including sessions in a decompression chamber, centrifuge runs, and flights aboard high altitude airplanes. These dogs were used for six flights during 1951 and 1952. Korolev flew two dogs at a time, and three of the animals made two flights each. Physiological measurements included pulse, respiration, and temperature. An onboard movie camera recorded their behavior. By the end of 1952 Korolev completed 18 biological flights involving 12 different animals.

Overall these flights went well, but sometimes events did not go as planned. The night before the sixth biological launch Bobik, one of the dogs that was to make the flight, escaped during her evening walk. Everyone tried to find her, but she had disappeared in the night. Rather than postpone the flight, the technicians caught one of the stray animals usually found loitering behind the kitchen. The next day they prepared the substitute for flight, and loaded her in the rocket. They named this animal ZIB, a Russian acronym for "Substitute for Bobik." A few minutes before launch Bobik ambled back into the blockhouse area, but it was too late to substitute her for ZIB. Despite her lack of preparation ZIB performed well during the flight.

Korolev continued launching the R-2s from Kapustin Yar, and his NII-88 team designed larger missiles in the early '50s. Although they were still being asked to design new vehicles, the Germans at Gorodomlya were completely out of the main stream of Soviet rocket development by this time. In 1953 Soviet authorities ordered their repatriation to their home country.

Liftoff of the third Operation Backfire V-2 on October 15, 1945. Observers from the United States and Russia were invited to witness the launch, along with members of the press. *Source: Mitchell R. Sharpe.*

11

Hermes and White Sands

Recognizing the importance of rockets in any future conflict, the United States Army initiated Project Hermes on November 15, 1944, for research and development of ballistic missiles. During December it was decided to study the V-2 and conduct live launchings as part of Hermes. At the time the largest American rocket being produced was the 16-foot tall WAC Corporal. Designed by Frank Malina, it carried a 25-pound payload to 100,000 feet. Hermes would unlock the secrets of German rocket programs, and provide potentially useful information for the fledgling American efforts. The General Electric Company received the Hermes contract.

Robert Goddard, the American rocket pioneer who launched the world's first liquid fuel rocket, spent most of World War II in Annapolis, Maryland, working on rocket motors to help heavily loaded seaplanes take off. In the weeks before his death from throat cancer on August 9, 1945, he examined components from the V-2 brought to America. One can only imagine the results if he had been given the same resources as the Germans.

In July 1945 the American Army embarked on Operation Overcast, a program to recruit up to 350 German scientists and engineers to work under contract in the United States. As originally conceived, Overcast was a temporary measure to exploit the talents of the Germans for the prosecution of the war against Japan. Under the terms of Overcast, German scientific and technical specialists would be offered three month contracts that were renewable for an additional nine months at the discretion of the United States Government. There were no provisions for the families of the Overcast personnel, nor were there any offers of eventual citizenship—situations they found unacceptable. Negotiations continued after Japan's surrender, during which time the area where dependents lived became known as "Camp Overcast." Since Overcast was supposed to be classified, the American military renamed the effort "Paperclip" in March 1946.

The name "Paperclip" came from the manner in which the personnel folders of the Germans identified for eventual recruitment were marked. A paperclip was placed in their folder as an identifier.

Project Paperclip is most often associated with the German rocket scientists, but it went much farther. Virtually every area of

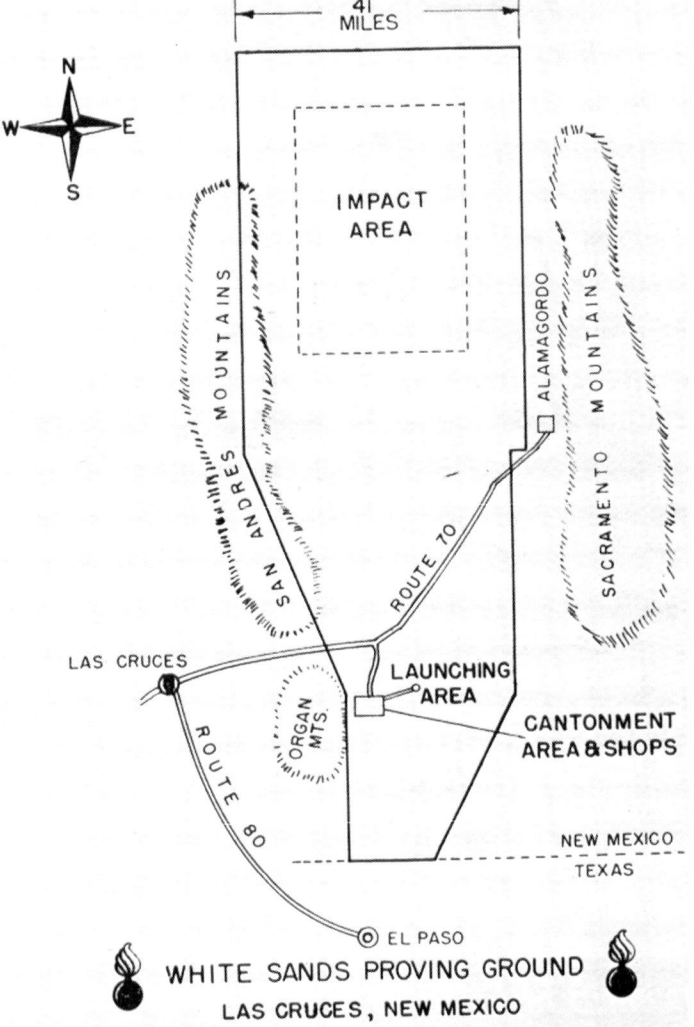

Established in 1945, the White Sands Proving Ground covered an expanse of land 41 miles wide and 100 miles long.

Chapter 11: Hermes and White Sands

Frank Malina stands alongside the WAC Corporal. In 1945 this was the largest American rocket being produced, and could carry a 25-pound payload to 100,000 feet. *Source: White Sands Missile Range Office of Public Affairs.*

German science, industry, and technology was examined by the American military for exploitation and possible application for the continuing war with Japan. After interrogation, researchers who had desired expertise were offered contracts to work in the United States. From the beginning the rocket specialists received different treatment from other German personnel. The rocketeers received six month contracts under Paperclip, while the standard contract in other areas was three months.

Launching captured V-2s required a large expanse of land, particularly if the rockets' impacts were to be observed, or payloads recovered. The Ordnance Department decided to launch the V-2s from the newly established White Sands Proving Ground in southern New Mexico. When creating White Sands Proving Ground, the Army Ordnance Branch used many of the same criteria Dr. Goddard did more than 10 years earlier. They sought a flat, sparsely populated area with predominantly clear skies. Having an area surrounded by mountains for tracking stations was a plus. The site also had to be alongside existing railroad tracks, and preferably near an established military base for support. In February 1945 the Army Corps of Engineers selected a locale adjacent to the White Sands National Monument, just outside of Alamogordo, New Mexico. The proving ground was named for the National Monument. Fort Bliss and the city of El Paso, Texas, were nearby, as was the New Mexico College of Agriculture and Mechanic Arts (today known as New Mexico State University) at Las Cruces.

Procuring the land was not a problem, since the American government already had leases on much of the land. The Corps of Engineers issued Real Estate Directive 4279 on February 8, 1945, which declared the use of the land to be of military necessity. The tract included the Alamogordo Bombing Range, Fort Bliss Antiaircraft Firing Range, Doña Ana Target Range, Castner Target Range, and the Jornada Experimental Range. Department of the Interior Special Use Permit Number 30, dated April 30, 1945, covered the use of the White Sands National Monument. One major highway,

L. B. Carter, of the General Electric Company, inspects combustion chambers shipped to White Sands from Mittelwerk. Missile components arrived in New Mexico in August 1945. To move the hardware from the railroad sidings, the Army hired every flatbed truck in Doña Ana County.

Chapter 11: Hermes and White Sands

U.S. Route 70, crossed the southern end of the range. The launch area was south of the highway; the impact area to its north. In the interests of public safety, the New Mexico Highway Department agreed to let the Army close the road during missile firings. (To this day, Highway 70 is still subject to hour-long closures due to testing. Travelers are advised to call (505) 678-1178 in Las Cruces or (505) 443-7199 in Alamogordo before driving across the range.)

Construction began on 25 June. Headquarters for the new testing center was near the foothills of the Organ Mountains, across the Tularosa Basin from Alamogordo. At the time the range was believed to be a temporary installation, so several old Civilian Conservation Corps structures were moved to White Sands from Sandia Base, near Albuquerque. The Army built a large hangar in the Industrial Area for assembly and preparation of V-2s; the launch site was about six and a half miles away.

Despite the planned temporary nature of the establishment, the headquarters area was divided into four areas: Administration and Troop; Technical; Industrial and Warehouse; and Quarters and Parade Ground. This was done in such a manner to allow future expansion and growth. On July 9, 1945, the Army officially established the White Sands Proving Ground (WSPG). One week later, on 16 July, scientists working on the Manhattan Project detonated the world's first atomic bomb at Trinity Site, which was within the confines of WSPG. (However, it should be noted there was no formal link between White Sands missile project and the Manhattan Project.)

Water was a major concern, as would be expected for an installation in the New Mexico desert, so the first construction project was the drilling of wells. Six wells were drilled to the south and east of the headquarters area. The engineers found several abandoned mines in the Organ Mountains that had filled with water. Although this water was not potable, it sufficed for construction purposes during the early months of the base.

WSPG was approximately 100 miles long and 40 miles wide. This made it the largest overland rocket and guided missile testing center in the United States. There were a few isolated ranches scattered across the range. At first, the Army signed contracts with the ranchers for the use of their land. During firings the ranchers had to be evacuated for their own safety. Within a few years, though, this proved increasingly unworkable, as the ranchers began ignoring the evacuation orders. The Army therefore acquired all the land within the confines of WSPG.

Rocket launchings began at White Sands on September 26, 1945, with the firing of a Tiny Tim rocket modified for use as a booster for the WAC Corporal sounding rocket. A two-stage WAC Corporal launched on 11 October reached 43.5 miles. At the time this was a record for an American rocket.

The WAC Corporal sustainer burned red fuming nitric acid and aniline, propellants that ignite spontaneously on contact. Such combinations are called *hypergolic* propellants. Pressurized nitrogen forced the propellants from their tanks into the combustion chamber. The rocket was fin stabilized, and did not have an active guidance system. Many people assumed the name of the rocket (WAC) was a tribute to the Women's Army Corps. Actually, it was an acronym for the manner in which the rocket flew, Without Attitude Control.

Of the more than 400 German missile scientists who surrendered to the American Army, just over 100 received contracts under Operation Paperclip. Under the leadership of Wernher von Braun, they arrived in the United States aboard the ocean liner *Argentina* on November 17, 1945. At first, the group was housed at Fort Strong, near Boston, Massachusetts. In January 1946 they were brought to Fort Bliss to assist with V-2 launchings and answer questions by Army and General Electric personnel.

By the time the German rocket scientists arrived at Fort Bliss, the missile components recovered from Mittelwerk were already there. When the Americans occupied the underground factory there were about 250 rockets on the assembly line in various degrees of completion; in other words, no complete rockets were found. In the few weeks remaining before turning the area over to the Russians, more than 640 tons of equipment requiring 300 railroad cars were shipped from Mittelwerk. This technological cornucopia included 215 combustion chambers, 180 sets of propellant tanks, 90 tail units, 100 sets of graphite vanes, and 200 turbopumps.

In August 1945 the missile components arrived at White Sands. To give some idea of the magnitude of the effort, every railroad siding from El Paso, TX, to Belen, NM—a distance of 210 miles—was full of cars bearing rocket hardware. The Army hired every flatbed truck in Doña Ana County to move the material. The task was completed in 20 days.

When the rocket components were inventoried, it was quickly discovered the Americans at Nordhausen, many of whom had no background in rocketry, had gathered whatever looked important. This resulted in a mixed lot of parts, some more useful than others. For example, while control components were in short supply, there were two railroad cars full of rock wool insulation. Only 50 control gyroscopes had been shipped from Mittelwerk, most of which were in poor condition. Each rocket required two gyroscopes. Seventy electrical distribution panels had been shipped, but most of them were without wiring. Project Hermes would be plagued by some of the same parts shortages encountered by the British. Only two missiles could be assembled from originally matched parts from the factory. The others were put together from the variety of subassemblies and individual components.

Work proceeded quickly on the captured material. On March 3, 1946, the first American assembled V-2 was ready for firing at White Sands. This rocket did not fly; rather, it was bolted to the recently completed static test stand and fired for 57 seconds to observe the engine's performance. The Germans had directed construction of a 100,000-pound thrust static test stand that closely resembled one they had used at Peenemünde. Built into the side of the Organ Mountains, it comprised a heavy concrete shaft that was open at the bottom on the side away from the cliff.

The first flight took place with missile #2 on 16 April. Immediately after take off the rocket began flying erratically, lost a fin, and began to arc over. The engine was shut off by radio command after only 19 seconds of powered flight, and the missile, which still carried a nearly full load of fuel, crashed a short distance from the

On March 3, 1946, the first V-2 assembled at White Sands was static fired on a test stand built into the side of the Organ Mountains. *Source: U.S. Army photograph.*

Chapter 11: Hermes and White Sands

Launch preparations of an early White Sands V-2. Note the German ground equipment, including oxygen tankers, the Meilerwagen, and a Magirus ladder. The tower at the right of the image is for the WAC Corporal. *Source: U.S. Army photograph.*

Closeup of an early White Sands V-2. A German Magirus ladder is being used to reach the control compartment. A gantry and service structure would be added to the pad area later. *Source: New Mexico Museum of Space History.*

launch pad. One of the graphite jet vanes broke off soon after ignition, and the rudder on the tip of the fin tried to compensate. This placed unusually large loads on the fin.

Launch personnel watched from the relative safety of a concrete blockhouse designed by Dr. Del Sasso, a group of CalTech engineers, and White Sands Proving Ground's first commander, Lieutenant Colonel Harold Turner. Construction of the blockhouse began on July 10, 1945, and was completed in September at a cost of $36,000. It was a squat building with a pyramid-shaped roof. Built of reinforced concrete, it had 10-foot thick walls; the roof was 27 feet thick at its apex, which made it (hopefully) capable of withstanding the impact of a V-2 falling from an altitude of 100 miles. The floor was 12 feet thick, just in case an errant missile burrowed into the ground.

The blockhouse consisted of a 937-square foot firing control room housing firing controls, monitoring and communications equipment, and test personnel. It had three viewing ports made of blast proof safety glass, a blast proof door, and a roof wash-down system designed to decontaminate the building in the event of a rocket explosion. The room was mechanically air conditioned and electrically heated.

Early V-2 launches from White Sands relied on German equipment like the Meilerwagen and Magirus ladder for servicing and preparation. To better service the V-2s, in December 1945 the Army decided to add a gantry crane that could accommodate rockets up to 54 feet high to the launch pad. Construction of the gantry began in August 1946. It was completed in November at a cost of $38,000.

The finished structure stood 63 feet tall, and was 28 feet wide. It consisted of two 60-foot open steel towers tied together at the top by an open steel truss. Three pairs of adjustable work platforms

To provide better access to the V-2s, the Army built a gantry and service structure at the pad. During launch it could be moved back from the rocket. *Source: U.S. Army photograph.*

Even without an explosive warhead, V-2s still made sizable craters when they hit the desert. The combustion chamber of this particular missile is visible near the bottom of the crater. *Source: U.S. Army photograph.*

Chapter 11: Hermes and White Sands

This view shows the blockhouse, a V-2 on the pad, and the service structure rolled back for a launch. The WAC Corporal launch tower is behind the V-2.
Source: U.S. Army photograph.

Cornelius X. Ryan, an employee of the General Electric company, at the launch control panel inside the V-2 blockhouse.

between the towers could be swung down to encircle a rocket and, for rockets higher than 54 feet, two outrigger platforms could provide access to one of the rocket's sides. The platforms were reached by stairs and ladders, and the top truss had a 15-ton chain hoist.

The gantry was equipped with fire fighting, communications, low pressure air, electrical service, and propellant handling equipment. For launch, the gantry was moved well away from the rocket. It rested on four pairs of 400-pound rail wheels, each connected to electric drive motors that moved the structure at a slow pace up and down the crane's 500-foot track.

Three weeks after the unsuccessful flight of the second White Sands V-2, missile #3 flew successfully. This missile reached an altitude of 70 miles after an engine burn of 59 seconds.

Captured German missiles provided the first opportunities for American scientists to launch payloads into space. If launched straight up, the V-2 could reach an altitude of more than 100 miles. Hermes V-2 flights did not carry explosives. However, because the V-2s were designed to carry one-ton warheads, they needed ballast to compensate for the missing explosives to remain stable. The Army reasoned if the rockets had to carry weight in their nose cones, why not let that weight be scientific instruments? Scientists leapt at the chance.

In January 1946 the Army Ordnance Department held a conference for military and university personnel to discuss the possibility of placing instruments on V-2s. Attendees included representatives of the Rocket Sonde Research Section of the Naval Research Laboratory, and the Applied Physics Laboratory (APL) of Johns Hopkins University. At the conference, it was explained that the Army's V-2 program had three goals: to gain experience handling and launching large liquid fuel rockets; to obtain ballistic data; and to make measurements of the upper atmosphere. Lieutenant Colonel J.G. Bain of the Ordnance Department extended a special invitation to the Rocket Sonde Research Section to participate in the upper atmosphere research program.

Hermes V-2s #2 and #3 carried rudimentary cosmic ray experiments in modified German warheads. The nose of each rocket contained a single Geiger tube shielded by an inch-thick lead cylinder. Power for the tube was supplied by an assembly of 52 22.5 volt hearing aid batteries set in wax. In addition to the Geiger tube, each rocket carried a roll of 35-millimeter film. The film would record the passage of radiation particles through the warhead. A gravity switch, set to function when the rocket reached an acceleration of 4 g's, was used to turn on the flight data recorder.

Subsequent missiles carried specially prepared plates of photographic emulsion. These often recorded the tracks of cosmic radiation particles. Similar emulsion packets were also carried aloft on high altitude balloons, which provided a longer exposure time due to their flight profile.

Cosmic radiation was an area of great interest at the time. Mountain climbers and balloonists carrying Geiger counters had previously noticed increases in radiation as they ascended. The source of this radiation was determined to be from outer space, hence the term cosmic radiation. Comprised mainly of high speed charged particles, nobody was sure how much of a health risk they posed to future space travelers, or even high altitude aircraft pilots.

Rocket #2 did not reach a sufficient acceleration to turn on the data recorder, so even though the warhead was recovered, there was no data. It is also unlikely that the instruments would have recorded any cosmic radiation anyway, because this rocket only reached an altitude of five miles. Rocket #3 created a large crater on impact, and no trace of the warhead was ever found.

The German warheads were generally unsuitable for scientific work, as the equipment was inaccessible once it was placed inside them. The Naval Gun Factory in Washington, D.C., manufactured new nose sections that were better suited for scientific use. They were fabricated from 1/8 inch thick cast steel, and contained a volume of 19.6 cubic feet. Three gasketed ports allowed access to the payload before launch. A pressure of one atmosphere was maintained inside the nose cone throughout the flight. The overall length of the new warhead sections was 7 feet, 6 inches, and they weighed 1,055 pounds.

Numerous other components, such as gyroscopes and electrical junction panels, were built in the United States for the Hermes V-2s. In some cases, the Americans improved upon the German design. Wiring throughout the missiles proved problematical. The Germans had used single strand copper wire, which broke easily. By 1947 General Electric personnel replaced the original single strand copper wire with more durable stranded wire. Missile com-

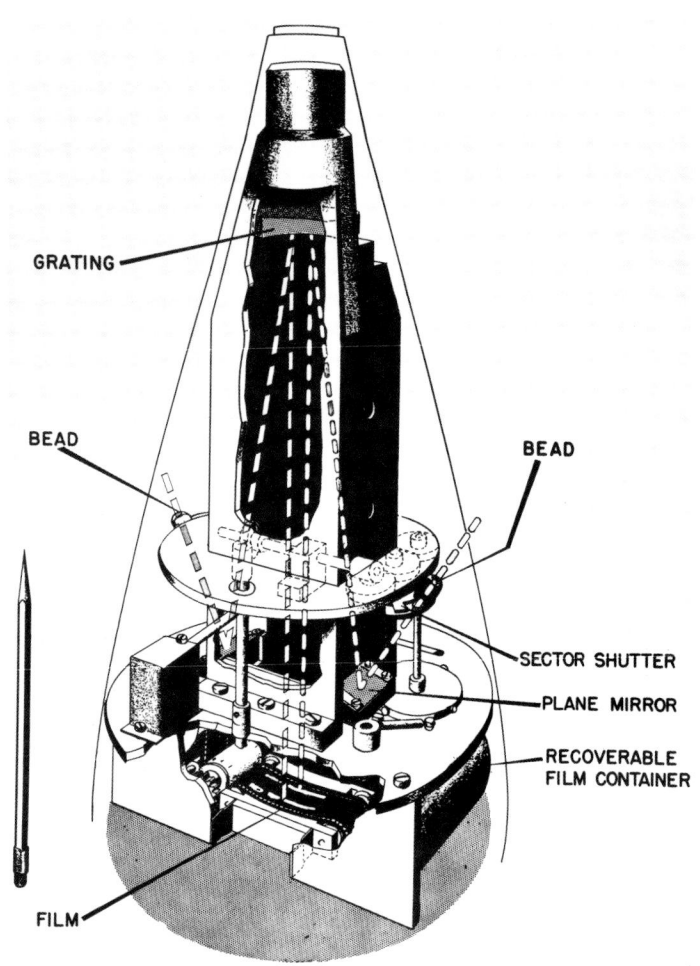

Solar spectrograph flown on the V-2 #12 at White Sands.

Chapter 11: Hermes and White Sands

Liftoff from White Sands. *Source: U.S. Army photograph.*

ponents also had to be thoroughly inspected, cleaned, and tested before use.

As evidenced by V-2 #3, payload recovery was another area that needed work. Even without the explosive in the warhead, the missiles fired at White Sands made craters nearly as big as those during wartime that hit Allied targets. Telemetry was in its infancy, so the instruments used on-board recorders. This meant the nose sections had to be recovered. To enhance the probability that experiments could be recovered, it was suggested that the missile be separated into two aerodynamically unstable sections before it re-entered the atmosphere. If this were done, the individual pieces would impact at considerably slower velocities than that of an intact missile. Explosive charges were used to break the V-2 into two pieces. In theory, the pieces would tumble and fall slow enough for the instrument recordings to survive. This was very much a trial and error process.

Explosive separation was first attempted with V-2 #5. Lengths of primacord explosive were attached to the missile about a foot aft of the joint between the midsection and the tail. The explosives worked as planned, and cut through the skin. However, the plumbing that connected the propulsion unit to the midsection held the rocket together. It hit the desert more or less intact, and none of the experiments were recovered.

For the next missile, which flew on 28 June, one-pound blocks of TNT were placed on each of the four main struts of the control compartment. The TNT detonated according to plan, but the warhead did not separate. V-2 #9 (launched on 30 July) carried one pound blocks of TNT and nitrostarch on each of the struts. The nose separated, and the body impacted in the desert without leaving a crater. Apparently the explosives worked a little too well, for although the body was in surprisingly good shape, no trace of the nose was ever found.

For V-2 #12, two-pound blocks of TNT were used. This time the payload section separated and was found. On the next missile, only the badly punctured base plate of the warhead was found. The explosives were cut in half on V-2 #23. The payload was recovered, but the base plate was badly bent.

Scientists began considering placing their instruments elsewhere in the missile. As discovered with V-2 #9, without the nose section, the body of the rocket would tumble as it fell, and it struck the ground at a speed of only a few hundred miles per hour. V-2 #12 carried a solar spectrograph in a conical housing built into fin II. Two spectrographs had already been flown (in the nose sections), but had never been recovered. This time the instrument was recovered, and returned measurements of the solar spectrum in the ultraviolet wavelengths that are normally absorbed by the atmosphere. Damage to the instrument was so slight that it was flown again.

Seven missiles were assigned to the Air Force's Cambridge Research Laboratory. One of the main objectives of the flights was to develop parachute recovery methods, so the Air Force sponsored flights were given the name "Blossom." Working under contract to the Cambridge Research Laboratory, the Franklin Institute Laboratories for Research and Development modified the missiles. The midsections were lengthened by one caliber (about 65 inches), and they carried larger warhead sections. These modifications increased the weight of the Blossom V-2s substantially. A typical Hermes V-2 weighed about 8,800 pounds without propellants. The lightest Blossom V-2 weighed 9,781 pounds; the heaviest 10,683 pounds. These missiles were modified so they would break apart in the air, and have the nose sections return via parachute. Of the seven Blossom V-2s, four flew successfully.

The United States Navy also conducted two experiments with V-2s. The first, Operation Sandy, was a launch of a missile from a ship. Two flyable V-2s and one dummy missile were provided for

V-2s on the deck of the USS *Midway*. The Army provided one dummy and two live missiles to the U.S. Navy for launch from the deck of an aircraft carrier. *Source: Official U.S. Navy photograph.*

Chapter 11: Hermes and White Sands

Liftoff from the deck of the USS *Midway*. The rocket rose at an angle, and only reached an altitude of 12,000 feet. *Source: Official U.S. Navy photograph.*

Sandy. They were shipped from White Sands to Norfolk, Virginia, where they were loaded on the aircraft carrier USS *Midway*. Army personnel assembled the missiles, and trained the Navy launch crew. Navy Commander P.G. Holt directed the launch team.

On Saturday, September 6, 1947, one of the Sandy V-2s was launched from the deck of the *Midway* while it was several hundred miles off the East Coast of the United States. The launch itself was successful, but shortly after leaving the deck the missile went out of control, caught fire, and exploded at an altitude of 12,000 feet after a flight of six miles. Immediately after the launch the *Midway* was able to launch its aircraft. Since the objective of Sandy was to see whether or not a large missile could be fueled and launched from a ship that was underway, and if the ship could resume normal operations immediately after launch, Operation Sandy was termed a success.

The second Navy experiment was one of the most spectacular American V-2 tests. Dubbed Operation Pushover, it was exactly what its name implied. A fully fueled V-2 was toppled over at White Sands to obtain data for estimating how much damage such a mishap would cause aboard a ship.

Hermes V-2s carried a wide variety of experiments and payloads. Among the payloads lofted aboard the German missiles were biological packages, solar spectrographs, cosmic ray telescopes, and even experiments where rifle grenades were used to attempt to create artificial meteorites.

Operation Sandy V-2 in flight. *Source: Official U.S. Navy photograph.*

Biological rocket flights began in the United States on December 17, 1946, when researchers from the National Institutes of Health launched five Lucite cylinders containing fungus spores aboard White Sands V-2 #17. Scientists hoped the spores would show if cosmic radiation caused genetic mutations or other abnormalities. The missile reached an altitude of 116 miles, the highest attained by any Hermes launch. Unfortunately, the experiment was not recovered.

An experiment that involved corn seeds showed how frustrating early attempts at payload recovery were. Investigators from Harvard University provided packets of a special strain of seeds that when germinated would (hopefully) show any signs of genetic mutations caused by cosmic radiation. The first packet of seeds flew on V-2 #7 on July 9, 1946. As described previously, the payload was not recovered. A second packet of seeds flew on V-2 #8, launched 10 days later. This rocket exploded 28.5 seconds after lift off. There weren't any more of the special genetically controlled seeds left, but researchers figured after two successive failures, their chances of recovery were slight anyway. A package of ordinary seeds was purchased at a store in Las Cruces and placed aboard V-2 #9. As luck would have it, the missile reached an altitude of 104 miles and the seeds, located in pouches tied to the structure inside the body, were recovered. There is no record as to what, if any, results were achieved from these seeds.

Two months after the first biological V-2 experiment, Blossom I carried fruit flies, rye seeds, and cotton seeds in a cylinder that also contained a radar beacon, a device to measure parachute opening shock, and a camera. Launched on February 20, 1947, the rocket reached an altitude of 60 miles before it ejected the Blossom canister. The 14-foot diameter nylon ribbon parachute worked, and the payload landed intact. No radiation induced mutations or other changes appeared in the test subjects or their offspring.

Fruit flies and plants seemed to show no ill effects from rocket flights, but what about primates? Would the accelerations and vibrations encountered during a rocket flight be tolerable? What effects would weightlessness have? In April 1948 the Parachute Branch at Wright Field invited the Aero Medical Laboratory Acceleration Section to launch a monkey aboard Blossom III.

Dr. James P. Henry, who headed the Acceleration Section, supervised the experiment, while Captain David G. Simons assisted him. Henry was already well known within the aviation community as the developer of the partial pressure suit for high altitude flight. Simons, a 1946 graduate of Jefferson Medical College in Philadelphia, had requested an assignment in a research facility when he entered the Air Force on August 17, 1947. First Lieutenant Simons was assigned to the Aero Medical Laboratory Acceleration Section. This assignment let him combine medicine with his other long-time interest, electronics. Promoted to Captain in 1948, Simons designed and built electronic devices for the laboratory centrifuge.

One day Dr. Henry asked him, "Dave, do you think man will ever go to the moon?" In college, Simons had read articles about space travel, and answered it certainly was possible. Henry continued:

"Well, what would you think of having an opportunity to help us put a monkey in a captured V-2 rocket that would be exposed to about two minutes of weightlessness, and measure the physiological response to weightlessness?"

After Simons' enthusiastic and unqualified yes, Henry appointed him Project Officer for the experiment.

Blossom III had a special nose cone; one that had a fairing added that resembled the cockpit of the planned X-2 supersonic research airplane. The Air Force wanted to collect data on the aerodynamics of the shape, and test the airplane's separation system for pilot escape. Because the X-2 would travel at three times the speed of sound, the pilot could not safely bail out or eject in an emergency. To provide a system of pilot escape, the X-2 had a detachable nose with a drogue chute. Once the aircraft nose slowed enough, the pilot was supposed to jump clear and land beneath his personal parachute. Bell Aircraft, builders of the X-2, provided a nose cone separation system for Blossom III that was like the one planned for the supersonic airplane.

Henry and Simons had only two months to design and build the capsule. They selected a nine-pound American-born Rhesus monkey for their passenger. Someone in the Parachute Branch nicknamed the monkey "Albert," and the name stuck.

Because it was added to the flight so close to the launch, the capsule had to fit in space left over by other experiments, so it was oddly shaped. Its irregular shape made it difficult to fabricate, and led to numerous leaks when pressurized. Simons tried to patch the leaks by dabbing rubberized sealant over them. By the time Henry and Simons got the capsule to New Mexico, some of the rivets had loosened and let air leak out. More rubber sealant was needed. When pressurized Simons discovered the capsule bulged slightly, and it would not fit in the V-2 nose cone. Simons welded an aluminum strap to the inside of the capsule. The heat from the welding melted the sealant, and created a vile smelling sticky mess that had to be scraped off. The strap kept the capsule from bulging, but it aggravated another problem, namely that the capsule was not as large as the doctors would have liked. Albert had to be placed in a very awkward position, with his chin against his chest, to fit in the tiny capsule. Somehow, Simons and Henry finished the capsule in time for the launch, which was set for June 11, 1948.

The night before the flight Simons anesthetized Albert with Sodium Phenobarbital, and attached biosensors that would measure his pulse and respiration. As a further protective measure Simons injected Albert with Luminal, a muscle relaxant, to help him endure a hard landing. The respiration sensor, which was a mechanical lever sutured to Albert's chest, stopped working after Simons sealed the capsule, but the heart sensor continued to work. Responding to the anesthesia, Albert went from thoracic breathing to abdominal breathing, so the lever was in the wrong location.

When Dr. Simons climbed the launch tower to load the capsule in the nose of the missile, he noticed someone had written "Alas, poor Yorick, I knew him well" across one side of the rocket. After Simons loaded the capsule in the missile the heart sensor

Chapter 11: Hermes and White Sands

showed no activity. Either the sensor failed, or Albert had suffocated in the cramped confines of the capsule. Simons and Henry surmised the latter was the more likely possibility. In any event, Albert would not have survived the 37-mile high flight, because the parachute tore away from the nose cone.

The next attempt to launch a monkey aboard a V-2 took place on June 14, 1949, on Blossom IV-B. With a year to prepare for Albert II, Henry and Simons built an improved capsule and instrumentation system. The new capsule was a 36 by 12-inch diameter cylinder, and offered plenty of room for the passenger. The breathing sensor that had proven so balky on Albert I was completely revamped. Albert II wore a miniature oxygen mask. A heated wire inside the mask registered each breath the monkey took.

Henry and Simons also refined their test procedures. Two monkeys were selected for the flight, a primary and an alternate. For several weeks before the flight blood was drawn from both to provide a baseline from which to measure any effects of cosmic radiation. On the morning of the launch the doctors sealed Albert II in his capsule, and loaded it in the rocket by X-45 minutes. (The early White Sands launches used the term "X" to denote launch time.)

Albert II reached 83 miles. There was no telemetry, but heart and respiration rates were registered on an internal recorder, so recovery of the payload was critical. Unfortunately, the parachute again failed. The nose section created a crater 10 feet across and five feet deep. Some pieces were buried 12 feet, and only a few fragments could be identified. Fortunately for the scientists, those fragments included the precious pulse and breathing recordings. Albert survived the launch and weightless portions of the flight with no ill effects. During powered flight Albert's pulse rate slowed from 190 to 110 beats per minute, and his respiration went from 90 to 60 breaths per minute. Twenty seconds after motor burnout, as the rocket coasted to its apogee, the heart rate returned to 190, but the respiration rate rose only to 65 breaths per minute.

(L to R) John Adderson, Louis Padderson (both from the New Mexico School of Mines), Lieutenant Colonel Harold R. Turner (first Commander of White Sands Proving Ground), Dr. James Van Allen, and Arthur Coyne (both from Johns Hopkins University) discuss an experiment that used M-7 rifle grenades to create artificial meteorites. *Source: U.S. Army photograph.*

After Albert II, Simons left the Aero Medical Laboratory to attend the School of Aviation Medicine at Randolph Air Force Base, Texas. He was graduated from the Advanced Course in Aviation Medicine as a certified Flight Surgeon in 1950. His next assignment was at Yakota Air Force Base in Japan as a Flight Surgeon for the Far East Air Force.

Dr. Henry continued flying biological payloads aboard V-2s. Albert III's V-2 exploded in midair on September 16, 1949. Three months later, on 8 December, the parachute (again) failed when Albert IV flew. Results were the same as Albert II; the data recordings showed the monkey tolerated acceleration and weightlessness.

With only one Blossom V-2 remaining, Dr. Henry decided to try another line of research, and placed a mouse aboard the missile, which flew during the summer of 1950. No attempt was made to measure the rodent's physiological response to rocket flight. Instead, a camera recorded the mouse's reaction to weightlessness. The rocket reached 85 miles. As with the earlier primate flights the parachute failed, but the film survived because the camera was heavily armored. It showed the mouse retained "normal muscular coordination" throughout the weightless portion of the flight, and:

"...it no longer had a preference for any particular direction, and was as much at ease when inverted as when upright relative to the control starting position."

One of the more unusual series of experiments to be carried aboard the V-2 flew in late 1946, when attempts were made to create artificial micrometeorites using rifle grenades fired from the missiles. One of the principal investigators for this experiment was Dr. James Van Allen, from Johns Hopkins University. It was calculated that the velocity of a jet from a standard shaped-charge rifle grenade was comparable to the lower range of meteor velocities. If an M-7 grenade was fired from a missile and detonated at a high enough altitude, then scientists theorized it might produce artificial meteors. Since the mass, velocity, and composition of the matter ejected by the detonation was known, data from this experiment would help scientists in their studies of natural meteors. The experiment would also yield data on the physical properties of the upper atmosphere.

V-2 #12, launched on October 24, 1946, was the first in the meteorite experiment. Black powder charges were substituted for the grenades in order to test the ejection system. As a bonus, the puffs of smoke from the black powder charges would provide a means of observing high altitude winds. The charges detonated as planned, at altitudes of 100,000, 160,000, and 200,000 feet. However, the black powder did not produce the discrete puffs the scientists had hoped for; rather, it resulted in smoke streamers that indicated there were high winds aloft, but could not be used to quantify the conclusions.

Modified M-7 grenades were placed on V-2 #17, which was launched at 10:12 PM MST on December 17, 1946. One observer on the ground reported seeing a streak of light from the rocket, but he was the only person who saw anything, and this was suspect. Post flight analysis indicated the ejection mechanism probably failed; subsequent tests revealed the jet from an M-7 grenade was too weak for this application anyway.

In early 1946 Colonel Holgar N. Toftoy, Chief of the Research and Development Division, Office of the Chief of Ordnance, approved the suggestion of using a V-2 to boost a WAC Corporal. This would provide a two-stage rocket capable of reaching extreme altitudes, and would greatly increase the possibilities of upper atmosphere research. On June 20, 1947, the Bumper Program was inaugurated to investigate launching techniques for a two-stage missile and separation of the two stages at high velocity, to conduct limited investigation of high-speed high-altitude phenomena, and to attain velocities and altitudes higher than ever reached.

Overall responsibility for these missiles was given to the General Electric Company, and they were included in the Hermes Project. The Jet Propulsion Laboratory of the California Institute of Technology was assigned responsibility for the theoretical investigations, the design of the second stage, and basic design of the separation system. The Douglas Aircraft Company built the WAC Corporal and the special V-2 parts required.

The V-2 instrument section housed the guide-rails and expulsion cylinders used as a launcher for the WAC Corporal. These cylinders were activated by means of a compressed air bottle through a pressure reducer and a solenoid valve. This valve was activated by the final cut-off signal of the V-2, causing the fins of the WAC Corporal to slide out of the three slots in the upper part of this warhead launcher.

Eight of these missiles were assembled during the Bumper Program, and the first six were launched at White Sands Proving Ground. The first Bumper-WAC flew on May 13, 1948. This was the first large, two-stage rocket launched in the Western Hemisphere. This first combination rocket had a short duration, solid propellant motor propelling the second stage, and the WAC attained only slightly more speed and altitude than the V-2. The purpose of this flight was to test the separation system.

Bumper #2 was fired on August 19, 1948 and, like Bumper #1, contained only a partial charge. The velocity of the V-2 was about 10 percent below normal, but the steering was good. Up to 28 seconds the propulsion system performed normally, but at 33 seconds the turbine malfunctioned. The WAC separated at a lower velocity than intended, and the booster actually reached a higher altitude than the upper stage.

On September 30, 1948, the third missile was launched. The second stage used a liquid propellant with 32 seconds burning time. Operation of the V-2 was successful, but the second stage motor exploded just prior to separation. The fourth Bumper, like the third, used a liquid propellant with 32 seconds burning time for the second stage. The flight appeared normal at first, but a break in the alcohol piping resulted in an explosion in the tail section at 28.5 seconds. Bumper #4 went out of control, and crashed into the desert.

Despite the problems encountered in the Bumper flights, it was decided to press on with a full flight. Bumper 5, fired on February 24, 1949, was the first Bumper to be fired with a fully tanked second stage. This allowed 45 seconds burning time with the WAC. Thirty seconds after take-off the V-2 had attained a speed of 3,600 miles per hour, and the V-2 and the WAC Corporal separated. The WAC, with its power added to that of the V-2, attained a speed of

Lieutenant Colonel Harold R. Turner, first Commanding Officer of White Sands Proving Ground, installs M-7 rifle grenades aboard V-2 #12. Source: U.S. Army photograph.

Chapter 11: Hermes and White Sands

5,150 miles per hour, and an altitude of approximately 250 miles. This was the highest altitude ever reached by a man-made object.

The nose cone carried instruments to measure temperatures at extreme altitudes. In addition, the WAC carried telemetry that transmitted technical data back to the ground. This was the first time radio equipment had ever operated at such extreme altitudes. Although the missile had been tracked by radar for most of its flight, more than a year passed before the smashed body section was located.

The sixth V-2 WAC combination missile to be fired at White Sands Proving Ground was launched on April 21, 1949. This missile also had a fully tanked second stage, and it was hoped that the performance of Bumper 5 could be surpassed. The nose cone was instrumented to record data on cosmic radiation at altitudes greater than could be reached by other missiles. Performance was normal for 47.5 seconds, when the cut-off relay was operated by some malfunction in the control system. The WAC did not separate from the V-2. Excessive vibration, due to structural changes made to accommodate the WAC Corporal, most likely caused this failure, as well as the failures of missiles 2 and 4.

Bumper missiles 7 and 8 were shipped from White Sands Proving Ground to Florida by standard Army tractor and flatbed for firing at the Joint Long-Range Proving Ground. Since the V-2 missiles previously shipped to Norfolk, Virginia, for Operation Sandy had been damaged in transit, modifications were made in the shipping cradle, in that the rigid tail support was replaced by a partially inflated truck tire, which provided a non-rigid support for the tail. The Army vehicle was driven with extreme care, and the missiles arrived in excellent condition.

In general, the conventional V-2 ground equipment was used. The one major change was in the type of working platform used to service the upper levels of the missiles. The platforms were made up of standard commercial iron pipe scaffolding of the type commonly used by painters. These assemblies were mounted on casters. The scaffolds, extending to about 55 feet above the concrete pad, had sufficient strength and rigidity for the purpose.

The first attempt to launch Bumper 7 was unsuccessful due to moisture collected within the missile. It was necessary to return it to the hangar, where it was dried and rechecked. Two steps were taken to reduce the probability of further condensation troubles: (1) silicone grease was applied at vulnerable points; and (2) the loading sequence was reversed to load liquid oxygen after loading hydrogen peroxide. These measures proved adequate in two subsequent launchings.

Work went on with Bumper #8. Launched on July 24, 1950, this was the first missile firing from Cape Canaveral. Bumper #7 was successfully fired on July 29, 1950. The experiments to be carried out on these missiles called for a relatively low trajectory, with a separation angle of approximately 20 degrees from horizontal. The General Electric Report on these firings stated:

"This trajectory required a relatively rapid turn during the powered flight of the V-2. Both missiles made the turn successfully, and the general performance appeared good. A closer examination of the trajectory data showed, however, that the program had been

Liftoff of Bumper 7 from Cape Canaveral, FL. General Electric launched eight of these two-stage rockets. The V-2/WAC Corporal combination was the world's first two-stage liquid fuel rocket. Bumper 5 set an altitude record of 248 miles at White Sands; Bumpers 7 and 8 were the first rockers launched from Cape Canaveral, FL. *Source: U.S. Army photograph.*

The V-2 booster for Bumper 7 prior to adding the WAC Corporal upper stage. Source: *U.S. Army photograph.*

Bumper 8 launch preparations. Due to a technical problem with Bumper 7 this was the first rocket launched from Cape Canaveral, FL. *Source: U.S. Army photograph.*

Chapter 11: Hermes and White Sands

greater than desired. Trajectory data showed the separation angle for Bumper 7 to be approximately 10 degrees, and that for Bumper 8 to be about 13 degrees. The fact that the two trajectories showed the same type of discrepancy indicated a systematic, rather than a random fault. Since it seemed highly improbable that the pitch device itself would fail in such a fashion as to increase the program, precession of the pitch gyro circuits had been modified to obtain a much larger than normal program, these circuits were among the first investigated. This investigation turned up a 'sneak-circuit,' which caused the erecting motors of the pitch gyro to be energized after take-off. This in turn caused a procession which operated to increase the program angle. This fault appeared to answer fully the observed discrepancy."

Despite the problems with the trajectory, Bumper 7 attained a speed of Mach 9, the highest sustained speed that had ever been reached in the Earth's atmosphere.

Hermes is most frequently associated with the V-2 firings, but it was actually the name for a much larger effort, and involved several different rockets. Over the years this has caused some confusion. Adding to the confusion, the missiles bore the designation Hermes. Some were anti-aircraft missiles, some were surface to surface.

Hermes II on the launch pad. *Source: U.S. Army photograph, courtesy of White Sands Missile Range.*

One of the most promising missiles developed at Peenemünde during the war was the *Wasserfall* (Waterfall) anti-aircraft rocket. Looking very much like a half-scale V-2 with short, stubby wings in the midsection, the Germans tested about 35 *Wasserfalls* by the end of the war. General Electric engineers copied the *Wasserfall* aerodynamic design, and designated it the Hermes A-1. Six were built; the first was damaged beyond repair during a static firing, while the remaining five flew at White Sands in 1950 and 1951. V-2 #19, launched on January 23, 1947, tested the telemetry unit intended for the Hermes A-1. Other V-2s launched in 1947 and 1948 carried Hermes A-1 telemetry and guidance equipment.

The Hermes A-1E1 and A-1E2 were both planned as improvements on the A-1 that were suitable for field use. During May 1950 the emphasis of Project Hermes shifted, so that the primary emphasis was on surface to surface missiles, not anti-aircraft, so work ended on both projects. Similarly, the Hermes A-2 did not progress beyond the drawing board at first. Then, in December 1948 the Hermes Project incorporated a study to develop an "inexpensive guided missile." This received the A-2 designation before it finally became the solid fuel RV-A-10. The Jet Propulsion Laboratory worked on the RV-A-10 project, which included hundreds of solid fuel motor tests.

In June 1946 the General Electric Company received another contract amendment for a missile dubbed Hermes II, a test vehicle for a two-stage anti-aircraft missile. As envisioned, the second stage of the Hermes II was a ramjet propelled vehicle capable of reaching an altitude of 66,000 feet, and a speed of 2,100 miles per hour. A V-2 boosted the second stage.

The first Hermes II test, missile 0, was launched on May 29, 1947. This was a standard V-2, with the proposed Hermes II guidance system. Four seconds after launch the missile was supposed to head north at an angle of 7° off the vertical. At first no pitch was evident, then observers at the emergency cut-off station realized it was headed slightly south. They judged the angle was so slight that

The Hermes II being loaded on the Meilerwagen in the Industrial Area of White Sands Proving Grounds. This view shows the enlarged fins and air rudders added to this missile. *Source: US Army photograph, courtesy of White Sands Missile Range.*

Chapter 11: Hermes and White Sands

the missile would still land within the confines of White Sands Proving Ground, so they let it continue on its course. As the missile flew, they realized it would overshoot the range. They feared if they shut off the engine it might hit the City of El Paso, so they let it continue. Cut-off occurred 46 seconds after launch, which was enough to take it past El Paso, across the Rio Grande River, and into Mexico. The rocket hit just outside Ciudad Juārez. Fortunately, there were no injuries.

Missile #44, launched on November 18, 1948, carried a ramjet diffuser in its nose. Aerodynamic data agreed with wind tunnel and theoretical studies as the missile accelerated to Mach 3.6. This rocket topped 90 miles altitude.

Hermes II missile #1 flew on January 13, 1949. This rocket carried the large wings on the forward section. For Hermes II, these wings housed the ramjet propulsion system for the second stage. A cap over the leading edge of the wings protected the inlet, and reduced drag during the ascending part of the trajectory. Small elevators were attached to the very front of the nose. Having the large wings near the nose of the V-2 affected its stability, so the fins were enlarged by 30%. Three Hermes II rockets flew. All carried dummy ramjet wings.

Hermes II #1 weighed 32,597 pounds at launch. The flight was postponed twice; once due to an onboard instrumentation problem, and again for weather. On 13 January the skies were overcast, but the winds were within the narrow limits allowed by the peculiar Hermes II configuration. The decision was made to launch regardless of the overcast. Radar and radio problems delayed the launch, but the missile finally lifted off at 1:26 PM. Twenty-two seconds after launch the rocket disappeared in the heavy clouds. At 33.4 seconds into the flight something happened, and the rocket broke apart. Debris from the shattered rocket landed about seven miles north of the launch pad, and were scattered over an area about a mile in diameter.

For the Operation Paperclip scientists this was a particularly frustrating time. They were housed at Fort Bliss, and most of them did not participate in the V-2 firings at White Sands. The number of Germans at White Sands reached its peak of 39 in March 1946. After that General Electric personnel replaced the Germans, and by the following spring the process had been completed.

General Electric's participation in the V-2 program ended on June 30, 1951, after 72 launches, counting the Hermes, Hermes II, Bumper, and Sandy flights. As Hermes progressed, scientists placed more and more instrumentation aboard the rockets, and began making modifications to the basic missile. In 1946 the White Sands V-2s weighed an average 150 pounds over their design weight, and none incorporated major contour modifications. By 1949 the missiles were weighing in at 1,036 pounds over their basic weight, and three quarters had design modifications. Owing to parts shortages, more and more of the missiles incorporated American built components. Douglas Aircraft even built eight tail sections for Hermes to replace ones that were too deteriorated for flight.

The end of Hermes was not the end of American V-2 flights, however. Throughout Hermes, the White Sands launches had been supported by the Army's 1st Guided Missile Battalion, which had been formed on October 11, 1945. After the Hermes V-2 flight program concluded, the 1st Guided Missile Battalion fired four rockets of their own. The first of these, TF-1, flew on August 22, 1951. Its engine fired until all the propellants were exhausted in a test to see how high an altitude could be attained by a six-year old German rocket launched by American soldiers. TF-1 reached an altitude of 132 miles, the highest attained by a V-2.

Missile #59, which was also identified as TF-2, flew on May 20, 1952, and climbed 64.3 miles. TF-3 reached 48.5 miles on August 22, and TF-4 never flew. The last American V-2 flight took place on September 19, 1952, when the 1st Guided Missile Battalion launched TF-5.

Officially 68 percent of the V-2 flights were successful; that is, they flew with no malfunctions. Of the remaining 32 percent, though, many were still scientifically useful. As an example, one missile was classified as a failure because the steering mechanism mal-

Ignition of V-2 #55. This rocket exploded a few seconds later when the engine faltered, and the nose cone separation charge detonated. *Source: U.S. Army photograph.*

V-2 number 57 in the gantry. This was one of the last V-2s launched at White Sands.

functioned. Despite the malfunction, this particular missile reached an altitude of 99 miles, and returned excellent scientific results.

Overall, the percentage of fully successful flights compared to unsuccessful ones does not give a clear picture of the results of Hermes. For America, the V-2 provided the first experience in handling and firing large missiles, and laid the foundation for future rocket development. By taking advantage of the work done at Peenemünde, the United States saved at least a decade and billions of dollars that would have had to have been spent to reach the same level of technology.

12

To the Moon

By late 1949, rocket and missile projects under the aegis of the U.S. Army had begun to proliferate. General Electric was hard at work on Project Hermes, which included the V-2 launches, Bumper, Hermes II, and follow-on studies. The Guggenheim Aeronautical Laboratory at the California Institute of Technology (known as GALCIT) had formed the Jet Propulsion Laboratory, and was busy with the Private and Corporal rockets. Even Bell Telephone Laboratories of the Western Electric Company managed to get into the act with the Nike antiaircraft missile. There were also various missile-related projects at places like the Army's Aberdeen Proving Ground and Picatinny Arsenal. To coordinate the myriad activities, the Army consolidated their rocket programs at Redstone Arsenal in Huntsville, Alabama.

Huntsville, which billed itself as the "Watercress Capital of the World," was the site of a plant that produced chemical munitions during World War II. In early 1950 von Braun and most of the team that followed him from Peenemünde moved from Fort Bliss to Redstone Arsenal, where they set up the Ordnance Guided Missile Center.

In July 1950, less than a month after Communist North Korea invaded the Republic of Korea, the Army Chief of Ordnance asked the Ordnance Guided Missile Center to design a missile with a 500-mile range. The Army subsequently redefined its needs, and called for a more modest 200-mile range missile that could be launched by mobile field crews.

This missile was based on the General Electric study for the Hermes C-1, which was turned over to the Ordnance Guided Missile Center. The final missile would use an engine being developed by the Rocketdyne Division of North American Aviation, and had jet vanes and rudders like the V-2. (Interestingly, North American had previously replicated the V-2 engine. Although never "fired," the American copy was used for "cold flow" tests.) Significant advances over its predecessor included an inertial guidance system and detachable warhead. The new missile was 70 inches in diameter, and stood 69 feet tall. In its infancy, the missile was known variously as "Ursa," "Major," and "Hermes C1," before it was offi-

Redstone missile launch at White Sands. Technologically, the Redstone was an outgrowth of the V-2. *Source: U.S. Army photograph.*

cially named "Redstone," after the Arsenal. Chrysler Corporation received the contract to manufacture Redstone.

The first Redstone was launched on August 20, 1953, from Cape Canaveral, Florida. Due to problems in the guidance system, the first Redstone only flew 8,000 yards before it crashed. These problems were quickly fixed, and Redstone established a reputation for reliability, as 35 of the first 38 flights were totally successful. The Rocketdyne A-6 engine that propelled the missile burned liquid oxygen and alcohol to produce a thrust of 75,000 pounds.

During this time, von Braun became a well-known and highly visible advocate for space exploration. He contributed to a series of articles published in *Collier's* magazine, and to a pair of books published by Viking Press: *Across the Space Frontier* (1952), and *The Conquest of the Moon* (1953). Illustrated with paintings by Chesley Bonestell, Rolf Klep, and Fred Freeman, his books and articles helped gain public acceptance for what had recently been considered science fiction. He also worked with Walt Disney to produce programs, films, and books on the exploration of space.

In 1954 Frederick C. Durant III, a Director for the American Rocket Society, and current President of the International Astronautical Federation, helped arrange a meeting between von Braun and Commander George Hoover, of the Office of Naval Research, to discuss how to launch a satellite. Other participants included Alexander Satin, David Young, Dr. Fred L. Whipple, and Dr. S. Fred Singer. The group agreed that a Redstone topped with a cluster of Loki solid-fuel rockets developed by the Jet Propulsion Laboratory could place a scientific payload in a 200-mile high orbit. Their proposal became known as Project Orbiter.

The Navy advanced a competing satellite proposal, Project Vanguard. The Vanguard rocket was designed from the outset as a space launch vehicle, which appealed to President Eisenhower. He preferred to have the first American satellite launched by a purely research vehicle, rather than by a converted military missile. Vanguard received approval for development, and Orbiter was shelved.

By the mid-1950s von Braun and the group at Redstone Arsenal were busy working on another ballistic missile for the Army.

Redstone tail units at Redstone Arsenal, Alabama.

Chapter 12: To the Moon

Much larger than the Redstone, this rocket was named Jupiter. Because Jupiter would fly higher and farther than Redstone, the missile warhead needed protection from reentry heating. Rather than go to the expense of using full scale Jupiter rockets for nose cone reentry trials, von Braun suggested a more economical test vehicle. He proposed using a Redstone topped with solid rocket motors to push a model of the Jupiter nose through the atmosphere. This was essentially a three-stage version of the four-stage Project Orbiter booster. He named the test bed Jupiter-C, to take advantage of the priority rating assigned to the Jupiter missile.

The Jupiter-C had three stages. It comprised an elongated Redstone topped by two concentric groups of scaled down Sergeant (another Army tactical missile developed by the Jet Propulsion Laboratory) solid fuel rocket motors. The second stage comprised a ring of 11 motors; three motors nested inside the ring made up the third stage. Recognizing how close the three-stage Jupiter-C could come to orbital speed, President Eisenhower ordered Major General John Medaris, commander of the Army Ballistic Missile Agency, to travel to Cape Canaveral and personally verify there would be no "accidental" satellites launched. (The Army Ballistic Missile Agency was the successor to the Ordnance Guided Missile Center as the lead office for missile work.) Meanwhile, Vanguard experienced a frustrating series of technical delays and problems that were not atypical for any new program.

Then, on October 4, 1957, the Soviets launched Sputnik-1, the world's first artificial satellite, using a missile developed by Korolev. Sputnik circled the earth every 96 minutes in an elliptical orbit with a low point, or perigee, of 140 miles, and a high point, or apogee, of 587 miles. It was a highly polished 22.8-inch diameter sphere with 4 whip antennae and weighed 184 pounds. Within the hermetically sealed ball, a radio transmitter broadcast a series of pulses or beeps. Sputnik caused consternation, dismay, and even fear in Washington political circles. Some American politicians saw sinister overtones in the Soviet achievement. After all, a rocket that could launch a satellite into orbit could also lob a nuclear bomb at America. How had the Soviets managed to launch a satellite?

After the R-1 and R-2, Korolev continued building larger and larger rockets. In 1954 Soviet leaders gave Korolev authorization to develop an intercontinental missile. Korolev and his team studied various configurations, finally settling on a "rocket packet" that used a cluster of identical rockets that could be jettisoned during the ascent as their propellants were used up. This was the seventh long-range rocket designed by Korolev, so it became known as the R-7.

Korolev was ready to test the R-7 by the spring of 1957. On 15 May the first R-7 lifted off. The launch succeeded, but during ascent one of the boosters malfunctioned and destroyed the vehicle. The next R-7 also failed, and another rocket was removed from the launch pad due to a continuing series of malfunctions. Korolev's critics openly questioned his design, and he found himself in a very difficult position. Despite the criticism, he remained convinced the R-7 design was sound. The next launch, on 21 August, proved him right. The rocket reentered over Kamchatka, just as planned. With this success, the Communist Party Central Committee issued a resolution that Korolev launch an earth-orbiting satellite using the R-7.

Launch of the Jupiter-C carrying Explorer-1, the United States' first artificial satellite, on January 31, 1958. The Jupiter-C was derived from the Redstone missile. *Source: National Aeronautics and Space Administration photograph.*

Both the United States and Soviet Union had announced plans to launch satellites during the International Geophysical Year. The Soviets published a description of their satellite, even listing its frequencies, and how to pick up the signals. Most western observers did not take the Soviet announcement seriously, smug in their assumption that the Communists did not posses the technology for such a feat. When the 100th anniversary of Tsiolkovsky's birth passed without a satellite, the Soviet pronouncement was dismissed as so much hype and bombast. Unknown to the West, Korolev labored day and night with the satellite project. He was determined to be first.

Once Khruschev saw the worldwide reaction to the satellite, he directed Korolev to launch a second one as soon as possible. One month later Korolev launched Sputnik-2, which carried a dog into orbit. Sputnik-2 was even larger than the first satellite.

Part of Eisenhower's reticence about using a Redstone-derived vehicle to launch the first satellite is that he wanted to establish the right of free passage through space over potentially hostile territory. This could be important once reconnaissance satellites were developed. Legally, he felt it would be easier to establish such a precedent if the satellite were boosted with a purely research rocket, rather than a converted military one. The launch of two Soviet satellites rendered such considerations moot.

Finally, on November 8, 1957, the Secretary of Defense gave von Braun the order to launch a satellite. Von Braun put his team at the Redstone Arsenal in high gear. At first he said he could launch a satellite in 60 days, but superiors gave him 90 days to provide additional time to make sure the flight succeeded.

Vanguard was at last ready for its first orbital attempt on December 6, 1957. A few seconds after ignition the first stage engine faltered, and the entire vehicle toppled over in a fierce conflagration. The six-inch diameter test satellite, scorched and battered, lay on the ground near the pad. Newspaper reporters referred to the failed attempt as "Flopnik." The prestige of the United States rested on von Braun and the U.S. Army.

Just 84 days after receiving the go-ahead, von Braun had an orbital vehicle ready. The only difference between it and the last Jupiter-C flown was the addition of a single scaled-down Sergeant as a fourth stage, topped with an 11-pound instrument package. The instrument package and attached fourth stage motor were called "Explorer-1." This was the 29th Jupiter-C. At 10:48 PM EST, January 31, 1958, Explorer-1 blasted off from Cape Canaveral, Florida. For the next hour, everyone waited until tracking stations in California reported hearing Explorer-1. It was confirmed—the United States had successfully orbited a satellite!

Von Braun could not share the celebrations with his staff that night, because he was not at the Redstone Arsenal, nor was he even at the Cape. Rather, he was in Washington, D.C., along with Dr. James Van Allen (who, since Project Hermes, had moved from Johns Hopkins University to the State University of Iowa), and Dr. William Pickering from the Jet Propulsion Laboratory. Van Allen designed and built the instrument package. Pickering was Director of the Jet Propulsion Laboratory, which provided the upper stages that boosted the satellite into orbit. With word of the success, the three of them triumphantly held a duplicate of Explorer-1 over their heads before an audience of reporters. Explorer-1 carried radiation sensors that led to the discovery of two bands of radiation that encircle the Earth. These became known as the "Van Allen Radiation Belts."

The Jupiter was much larger than Redstone, with a range of 1,500 miles and a payload of 2,500 pounds. Like the V-2 and Redstone, it was launched by mobile field batteries. The first two, which were launched in 1957, failed. After these failures, though, a string of successful flights followed. Von Braun subsequently used the Jupiter as the first stage of a space launch vehicle that became known as the Juno II. Like the Jupiter-C (which von Braun called the Juno I), this rocket employed clusters of scaled-down Sergeant motors as upper stages. With the increased power of the Jupiter, though, the Juno II proved capable of launching probes to the Moon and beyond. On March 3, 1959, a Juno II blasted the Pioneer 4 spacecraft past the Moon. This was the first successful American lunar spacecraft. Like Explorer 1, Pioneer 4 was built at the Jet Propulsion Laboratory, and carried instruments provided by the State University of Iowa.

Sputnik sparked an overhaul of the American space program. The National Aeronautics and Space Act of 1958 created a new civilian space agency. NASA (for National Aeronautics and Space Administration) initially comprised the National Advisory Committee for Aeronautics (NACA) laboratories, the Naval Research Laboratory Vanguard team, and the Air Force large rocket engine programs. In December 1958 the Jet Propulsion Laboratory became part of NASA.

At that time, the Army Ballistic Missile Agency was not incorporated into NASA, and von Braun continued working on larger rockets under Army sponsorship. His next large rocket, dubbed "Saturn," comprised eight Redstone tank sections clustered around a Jupiter tank. Eight Jupiter engines propelled the rocket, which was then the largest ever built in the United States. Early in 1960 the Army Ballistic Missile Agency team working on the Saturn booster project was transferred from the Redstone Arsenal to NASA, creating the Marshall Space Flight Center. Von Braun became the center's first director.

Less than a week after the space agency's creation, NASA managers announced the establishment of a manned space project, which became known as Project Mercury. An office called the Space Task Group was set up at Langley Research Center in Hampton, Virginia, to manage Mercury. They selected the Air Force Atlas intercontinental ballistic missile to launch the Mercury capsules into orbit. Convair Astronautics of San Diego, California, built the Atlas. Kraft Ehricke, a former member of the Paperclip group, was one of the managers of the Atlas program for Convair. Prior to being assigned to the *Versuchskommando Nord* at Peenemünde, Ehricke had been a platoon leader in a *Panzer* unit on the Eastern Front.

NASA engineers realized many tests of the spacecraft could be made with suborbital flights using a smaller, less expensive rocket. The Space Task Group chose Redstone for this role. Before it could be used for manned flights the Redstone required some work. The military missile did not have a long enough thrust duration to boost Mercury on a suborbital flight. The ABMA had two Jupiter C rockets (which had longer tanks than a standard Redstone)

Chapter 12: To the Moon

The Saturn I was created by clustering eight Redstone tank sections around a Jupiter missile tank. It first flew in 1961. *Source: National Aeronautics and Space Administration photograph.*

on hand that they turned over to NASA. With longer tanks, the Jupiter C would burn for 143.5 seconds, 20 seconds longer than a standard Redstone. The Jupiter C normally burned a mixture of unsymmetrical diethyltriamine (UDETA) and liquid oxygen. For piloted flights, NASA replaced the UDETA with less toxic alcohol. Since the Rocketdyne A-6 engine was due to be replaced with the improved A-7 version, NASA opted to incorporate this engine on all the Mercury Redstone vehicles, rather than risk running short of hardware during the program. The A-7 generated a thrust of 78,000 pounds.

For Mercury, an 11-foot long instrument section was added to the top of the Jupiter C body. This proved problematical, because with the added section the Mercury Redstone was not as stable as the standard missile. Further analysis showed the Mercury Redstone combination would become unstable in supersonic flight, about 88 seconds after launch. To compensate for this NASA added 687 pounds of ballast to the instrument section. The ballast was a lead-filled plastic mixture that also helped damp out unwanted vibrations.

NASA also opted to use the LEV-3 autopilot for guidance, rather than the ST-80 Redstone stabilized platform. Although the LEV-3 was not as sophisticated as the ST-80, it was more reliable and would suffice.

Following four Mercury Redstone flights, including one carrying a chimpanzee named HAM, final preparations were underway for the first manned launch when Korolev again upstaged the Americans. On April 12, 1961, Red Air Force Major Yuri Gagarin orbited the Earth once in a spacecraft named Vostok.

Gagarin's successful flight did not dissuade NASA from their ordered, systematic approach leading to the first American orbital flight. On May 5, 1961, Commander Alan B. Shepard, Jr., became the first American in space. His suborbital flight lasted 15 minutes. An Army Redstone powered his capsule, which he had named Freedom-7. Less than three weeks after Shepard's flight, President John F. Kennedy issued a proclamation that America would send an astronaut to the Moon by the end of the decade. Project Apollo, which was on the drawing boards as an advanced successor to Mercury, became the American lunar effort.

Saturn made its debut on October 27, 1961, with a suborbital test of the first stage. Although it was the largest rocket in the United States' inventory, it still wasn't powerful enough for the lunar mission. Von Braun and the engineers at the Marshall Space Flight Center would have to build an even larger rocket, a behemoth that stood more than 36 stories tall and weighed 6 million pounds.

The Saturn, flown for the first time in 1961, became the Saturn I. It was followed five years later by the Saturn IB. Two more years passed, and the Saturn V lifted off for its first flight. The Saturn V took us to the Moon in July 1969. Many of the project managers on the Saturn V began their rocket careers at Peenemünde. For example, when the first lunar mission lifted off from the Kennedy

The Saturn IB, also sometimes referred to as the Uprated Saturn I, was the first of the Saturn series of rockets to carry humans into space. It comprised the Saturn I first stage, and a more powerful second stage than its predecessor. *Source: National Aeronautics and Space Administration photograph.*

Chapter 12: To the Moon

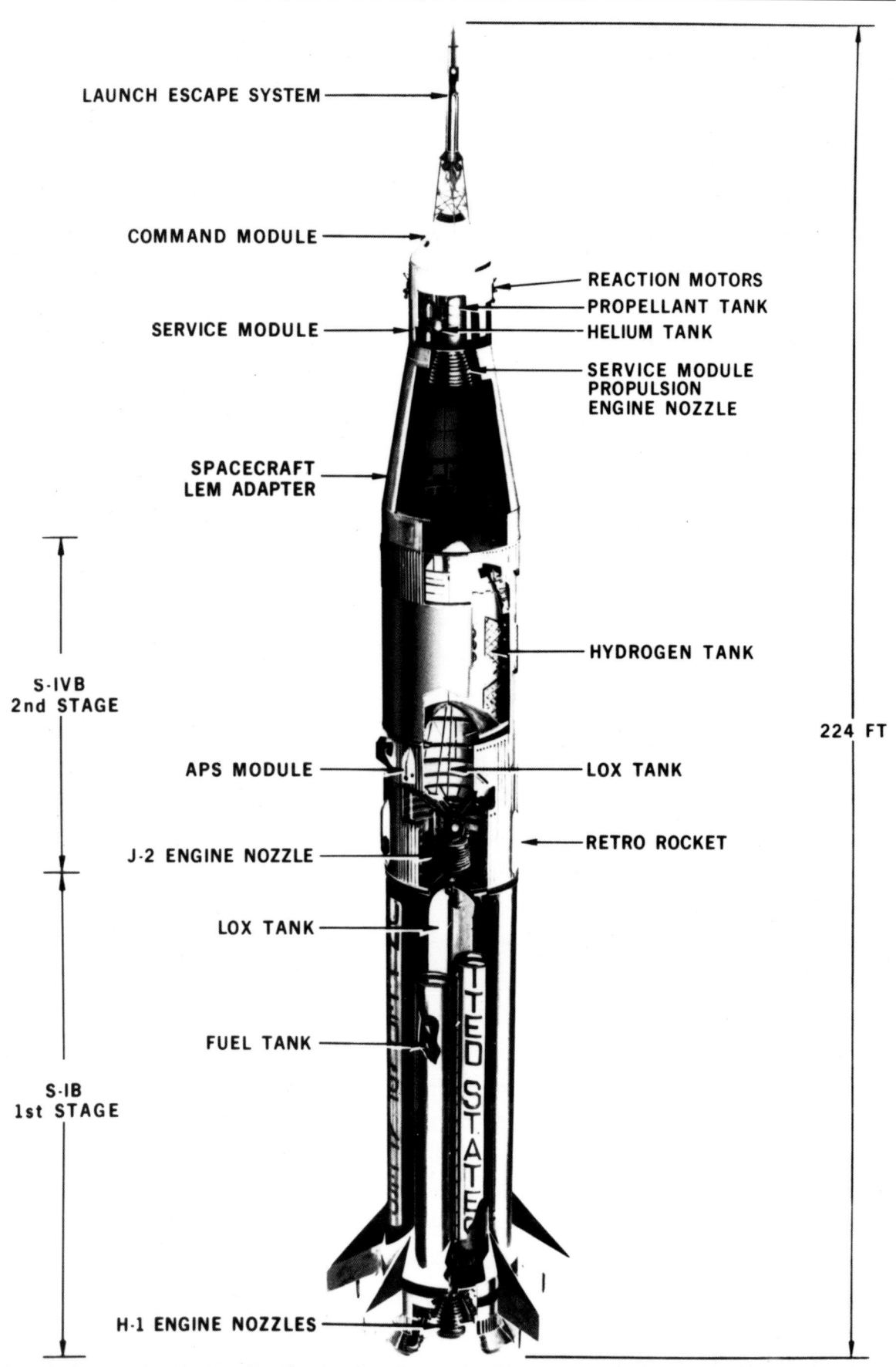

Saturn IB cutaway. Source: National Aeronautics and Space Administration photograph.

Chapter 12: To the Moon

Space Center, Dr. Kurt Debus was the Center Director. At Peenemünde he had managed the test stands.

In all, there were 13 Saturn V flights. Today, three remain in museums at Houston, Texas, Cape Canaveral, Florida, and fittingly, Huntsville, Alabama. (A fourth first stage is exhibited at Bay St. Louis, Mississippi, where it was assembled.) The Saturn V at Huntsville is in the rocket park at the U.S. Space and Rocket Center, along with Redstone and Saturn I rockets. With such gargantuan craft on display, it is easy to overlook the 46-foot tall rocket with 4 fins that also occupies a spot in the park. It is a V-2, grandfather of the mammoth rocket that took us to the Moon.

Opposite: Saturn V. On July 16, 1969, a Saturn V launched Neil Armstrong, Buzz Aldrin, and Michael Collins into space on the first lunar landing mission. The lineage of the Saturn V can be traced directly back to the V-2. *Source: National Aeronautics and Space Administration photograph.*

Appendix I:
American V-2 Firings

The information in the following table is arranged in chronological order, and includes White Sands Proving Ground launches, Project Bumper, Operation Sandy, and Hermes II flights.

V-2 #	Date	Altitude (Miles)	Remarks
1	March 15, 1946	n.a.	Static firing
2	April 16, 1946	3.4	
3	May 10, 1946	70	
4	May 29, 1946	69.7	
5	June 13, 1946	73	
6	June 28, 1946	67	
7	July 9, 1946	83.5	
8	July 19, 1946	3	
9	July 30, 1946	100.4	
10	August 15, 1946	4	
11	August 22, 1946	0	
12	October 10, 1946	108	
13	October 24, 1946	65	
14	November 7, 1946	0.2	
15	November 21, 1946	63	
16	December 5, 1946	95	
17	December 17, 1946	114	
18	January 10, 1947	72.2	
19	January 23, 1947	31	
20	February 20, 1947	68	
21	March 7, 1947	101	
22	April 1, 1947	80.3	
23	April 8, 1947	63.5	
24	April 17, 1947	88.5	
26	May 5, 1947	84	
0	May 29, 1947	49.3	Hermes II Missile 0
29	July 10, 1947	10	
30	July 29, 1947	99.9	
	September 6, 1947	6	Project Sandy Launch from USS *Midway*
27	October 9, 1947	97	
GE Spec.	November 20, 1947	16.6	
28	December 8, 1947	65	
34	January 22, 1948	99	
36	February 6, 1948	69	
39	March 19, 1948	3.4	
25	April 2, 1948	89.5	
38	April 19, 1948	34.8	
Bu-1	May 15, 1948	V-2 – 69.1 WAC – 79.1	Project Bumper
35	May 27, 1948	86.8	
37	June 11, 1948	38.7	
40	July 26, 1948	54	
43	August 5, 1948	103	

V-2 prior to launch at White Sands Proving Ground. *Source: U.S. Army photograph.*

Bu-2	August 19, 1948	V-2 – 8.28 WAC – 8.1	Project Bumper
33	September 2, 1948	93.6	
Bu-3	September 30, 1948	V-2 – 93.4 WAC – ***	Project Bumper
Bu-4	November 1, 1948	V-2 – 3 WAC – ***	Project Bumper
44	November 18, 1948	90.3	
42	December 9, 1948	67.4	
1	January 13, 1949		Hermes II Missile 1
45	January 28, 1949	37.2	
48	February 17, 1949	62.5	
Bu-5	February 24, 1949	V-2 – 63 WAC – 248	Project Bumper
41	March 21, 1949	83	
50	11 Apr 49	54.2	
Bu-6	21 Apr 49	V-2 – 31 WAC – ***	Project Bumper
46	May 5, 1949	5.5	
47	June 14, 1949	83	
32	September 16, 1949	2.6	
49	September 29, 1949	93.7	
2	October 6, 1949		Hermes II Missile 2
56	November 18, 1949	77	
31	December 8, 1949	81	
53	February 17, 1950	92.4	
Bu-8	July 24, 1950		Project Bumper; launch at Cape Canaveral, FL
Bu-7	July 29, 1950		Project Bumper; launch at Cape Canaveral, FL
51	August 31, 1950	84.8	
61	October 26, 1950	5	
2-A	November 9, 1950		Hermes II Missile 2-A
54	January 18, 1951	1.0	
57	March 8, 1951	1.9	
55	June 14, 1951	0	
52	June 28, 1951	3.6	
60	October 29, 1951	87.6	
TF-1	August 22, 1951	132.6	Training flight by 1st Guided Missile Battalion
59	May 20, 1952	64.3	Also identified as TF-2
TF-3	August 22, 1952	48.5	
TF-5	September 19, 1952	16.8	Last American V-2 firing

Appendix II:
The A-9/A-10 and A-4b

During his interrogation by the U.S. Army, von Braun revealed the details of a mammoth two-stage rocket; the A-9/A-10 combination, which would have been capable of reaching New York. As envisioned, the A-9 would have been a modified A-4 that fit into the nose of a mammoth rocket, the A-10.

The A-10 would have stood nearly 65 feet tall, and used an engine that generated a thrust of 441,000 pounds. Fifty seconds after liftoff, at an altitude of 80,000 feet and a speed of over 3,900 feet per second, the A-9 was to have begun firing. Its own engine would have boosted it across the Atlantic Ocean, over a range of 3,100 miles. This concept created quite a stir within the American military, for it was the first serious design for an intercontinental ballistic missile, and New York was its intended target. However, by 1943 Dornberger realized such a behemoth was not likely to be developed anytime soon, so he ordered it canceled.

Far more serious (and within the realm of possibility) was a proposal to launch missiles against the United States using U-boats. For this plan a V-2 would have been placed in a watertight canister, and towed horizontally behind a U-boat. Once they were within range of the target, ballast tanks in the base of the canister would have been flooded to tip the missile into an upright position. Then, the plan was to fuel and launch the missile against the American mainland. Fortunately, this plan was never put into action.

Late in the war, the idea of adding wings to an A-4 was revisited as a means to double the range of the missile. With this capability, missiles could have been launched against England from within Germany. In December 1944 the first such missile, the A-4b, was launched from Test Stand X at Peenemünde. The "b" reportedly stood for "bastard." This missile went out of control, and crashed shortly after launch. The second A-4b, launched on January 24, 1945, was far more successful. It reached a speed of Mach 4 and an altitude of 50 miles before it disintegrated.

The engineers at Peenemünde even created a design for a piloted A-4b. Instead of a warhead it carried a pressurized cabin, and had tricycle landing gear. Total flight time would have been 17 minutes, during which it would have traveled 400 miles. This design never left the drawing board.

Appendix III:
From Peenemünde to the Moon

1903	Konstantine Tsiolkovsky publishes theories about rocketry and space travel; proposes design for manned spacecraft.
1919	Robert H. Goddard publishes *A Method of Reaching Extreme Altitudes*.
1923	Hermann Oberth publishes *Die Rakete zu den Planetenraumen* (*The Rocket into Planetary Space*).
March 16, 1926	Robert H. Goddard launches world's first liquid fuel rocket.
1927	German *Verein für Raumschiffahrt* (Society for Space Travel or VfR) is formed.
December 17, 1930	German Ordnance Board appointed Walter Dornberger to head rocket development program.
February 21, 1931	Johannes Winkler launches liquid oxygen/liquid methane rocket; at the time believed by VfR to be world's first liquid fuel rocket.
November 1, 1932	Wernher von Braun begins working for German Army Ordnance Department at Kummersdorf.
August 17, 1933	Launch of GIRD-09 rocket; first hybrid rocket, liquid oxygen and "solidified benzene" fuel; first rocket using a liquid propellant in the Soviet Union.
November 25, 1933	Launch of GIRD-X, first all-liquid propellant rocket flown in Soviet Union.
December 1934	Successful flights by a pair of A-2 rockets.
April 1936	Overall design parameters for A-4 long-range rocket established.
December 3, 1937	First launch attempts of two A-3 rockets; guidance systems failed on both flights.
July 19, 1942	*Luftwaffe* contracts with Gerhard Fiesler Werk for construction of Fi-103 cruise missile.
October 3, 1942	First successful flight of German A-4 (V-2) rocket; first man-made object to leave the atmosphere.
May 26, 1943	German Long Range Bombardment Commission met at Peenemünde to determine which weapon, A-4 or Fi-103, should be deployed; group decided that both should be used.
August 17-18, 1943	Royal Air Force bombs Peenemünde under the code name "Operation Hydra."
November 5, 1943	Test launches of A-4 begin at Heidelager near Blizna, Poland.
May 18-20, 1944	Three-day exercise conducted at Heidelager to evaluate operational readiness of A-4.
June 13, 1944	Germans begin firing Fi-103 cruise missile against London; missile better known as V-1 "Buzz Bomb."
June 13, 1944	A-4 crashes in Sweden; Allies recover more than two tons of debris from the missile.
August 1, 1944	Heeres Versuchs Peenemünde becomes Electromechanische Werke, Karlshagen; effectively removed further A-4 development from Army control.

Appendix III: From Peenemünde to the Moon

August 8, 1944	*SS Brigadeführer* Hans Kammler appointed SS Commissioner General for the A-4 Program; deployment of missile taken away from Army and placed under Kammler's control.
September 6, 1944	First launch of V-2 in combat; missile directed at Paris; failed.
September 8, 1944	First successful combat launch of V-2; missile fired at Paris; later that day, first V-2 fired at London.
March 28, 1945	Last firing of a V-2 in combat.
May 8, 1945	World War II ended in Europe; more than 100 scientists and engineers who worked on German rocket programs under the leadership of Wernher von Braun were recruited by the U. S. War Department and brought to America to work on U. S. programs.
July 9, 1945	White Sands Proving Ground officially established.
September 26, 1945	First rocket launch at White Sands Proving Ground, a Tiny Tim modified for use as a booster for the WAC Corporal.
October 2, 1945	British launch V-2 from the Krupp Proving Ground near Cuxhaven under "Operation Backfire."
October 3, 1945	British launch V-2 from the Krupp Proving Ground near Cuxhaven.
October 15, 1945	British launch V-2 from the Krupp Proving Ground near Cuxhaven; code named "Clitterhouse," this was the last launch of Operation Backfire.
November 17, 1945	Wernher von Braun and more than 100 German missile scientists arrive in the United States.
March 3, 1946	First static firing of a V-2 at White Sands Proving Ground.
March 22, 1946	First American-designed rocket to reach space, WAC Corporal; climbed to 50 miles after launch from White Sands Proving Ground.
April 16, 1946	First flight of an American assembled V-2 at White Sands Proving Ground.
July 30, 1946	White Sands V-2 #9 carried packet of corn seeds; payload recovered.
February 20, 1947	Blossom I V-2 carried several vials of fruit flies, rye seeds and cotton seeds to an altitude of 60 miles; payload recovered intact.
September 6, 1947	Project Sandy launch; V-2 fired from deck of USS Midway.
October 18, 1947	First launch of a Soviet assembled V-2 rocket from Kapustin Yar.
June 11, 1948	Albert I, first monkey flight aboard V-2; monkey apparently died before launch.
September 17, 1948	First launch of Soviet R-1 (copy of German V-2) from Kapustin Yar.
February 24, 1949	Bumper 5 two-stage rocket; WAC Corporal upper stage reached 248 miles.
April 21, 1949	First launch of Soviet R-1A geophysical rocket from Kapustin Yar.
June 14, 1949	Albert II flight, second monkey launched aboard a V-2; instrument recordings indicated monkey was alive until impact; parachute failed.
September 16, 1949	Albert III flight with monkey onboard; V-2 exploded during launch.

Germany's V-2 Rocket

December 8, 1949	Albert IV flight with monkey aboard a V-2; instrument recordings indicated monkey was alive until impact following parachute failure.
July 24, 1950	Launch of Bumper 8; first rocket launch from Cape Canaveral, Florida.
August 31, 1950	Mouse launched aboard a V-2; mouse died on impact but camera survived landing, returning film of rodent's response to weightlessness.
July 22, 1951	Sergei Korolev begins biological rocket flights with dogs and other small animals at Kapustin Yar with R-1A "geo physical rocket."
August 22, 1951	Launch of TF-1 V-2 by the 1st Guided Missile Battalion at White Sands Proving Ground; first launch by an American Army crew.
September 19, 1952	Last American V-2 launch.
August 20, 1953	First launch of U. S. Army Redstone missile.
October 4, 1957	Soviet Union launches Sputnik-1; first artificial earth satellite.
January 31, 1958	Launch of Explorer-1, first American satellite by Jupiter-C; booster based on Redstone.
April 12, 1961	Launch of Vostok-1 by Soviet Union; Major Yuri Gagarin orbited the Earth once; first manned space flight.
May 5, 1961	Launch of Alan Shepard aboard Freedom-7 capsule by Redstone; first American manned space flight.
May 25, 1961	President John F. Kennedy announces the United States will "land a man on the moon and return him safely to the earth before this decade is out."
July 21, 1961	Mercury Redstone-4, Liberty Bell 7; suborbital flight piloted by Virgil I. "Gus" Grissom; pilot survived but spacecraft lost during recovery.
October 27, 1961	SA-1, first flight of the Saturn-1 launch vehicle; suborbital test with dummy upper stage.
February 27, 1966	First flight of Saturn IB.
November 9, 1967	Apollo-4; first flight by Saturn V with unmanned spacecraft.
October 11-22, 1968	Apollo-7; first manned flight launched by a Saturn IB booster.
July 16-24, 1969	Apollo 11; first manned lunar landing; launched by a Saturn V.

Appendix IV:
The Canadian Arrow

Rockets resembling the V-2 may once again fly. These rockets, however, will be built by the Canadian firm PlanetSpace, and will carry passengers. Called the Canadian Arrow, these modern rockets are based on the V-2. They will carry up to three people at a time to altitudes in excess of 100 kilometers.

The Canadian Arrow was PlanetSpace's contender in the Ansari X-Prize competition. The X-Prize was a $10-million award offered to the first non-government team that could build and launch a spacecraft carrying three humans to a suborbital altitude of 100 kilometers on two consecutive flights within two weeks. The rules allowed ballast to be substituted for two of the occupants. On October 4, 2004, Spaceship One, built by Burt Rutan and Scaled Composites, won the prize. However, this did not deter the PlanetSpace team from continuing their work.

PlanetSpace was the result of a merger between Geoff Sheerin, who was President of Canadian Arrow, and Dr. Chirinjeev Kathuria, previously of MirCorp—the organization that signed up Dennis Tito to be the world's first space tourist, or "citizen explorer," to the International Space Station. They expect PlanetSpace will fly 2,000 people into space in its first five years, with fares starting at $250,000. Sheerin, who designed the Canadian Arrow, based his rocket on the V-2 design. Both the nose cone and tail structure were identical to the German missile.

Standing 54 feet tall, the Canadian Arrow resembles a "stretched" V-2, making it look like the Soviet R-2. It is a two-stage vehicle. The first stage comprises a liquid-fuel booster that burns alcohol and liquid oxygen to generate a thrust of 57,000 pounds. The combustion chamber is based on the V-2 engine. One notable difference between the newer engine and its German ancestor, however, is that the Arrow's powerplant does not have a turbopump; gas pressure forces propellants into the combustion chamber. Four solid fuel motors boost the second stage, which houses the crew cabin, to suborbital altitudes. The solid fuel second stage motors also double as the launch escape system should something go awry during the initial boost. Both stages are parachute recovered.

PlanetSpace plans to begin flying the Canadian Arrow in 2007. After initial test flights, PlanetSpace officials expect to have decommissioned hardware available that they will use to build a full-size replica of the V-2 for donation to White Sands Missile Range. This rocket will be erected in the launch tower at Launch Complex 33, the original V-2 launch pad.

Bibliography

Armstrong, Clare H., et. al. *V-2 Rocket Attacks and Defense*. United States Forces, European Theater, Antiaircraft Artillery Section, Study No. 42, File No. R 471.6/1, N. D. (ca. 1946).

AVKO (Altenwalde Versuchs Kommando). *Die Fernrakete* (The Long Range Rocket). Document prepared by German personnel working on Operation Backfire, 1945.

Axelsson, George. "Nazis Keep Repeating Claims for V-2 Weapon," *The New York Times*, August 27, 1944.

Bilek, V. H., and McPhilmy, J. D. *Production and Disposition of German A-4 (V-2) Rockets (Project No. XT-1)*. Headquarters Air Materiel Command, Wright Field, Dayton, Ohio: Staff Study No. A-SS-2167-ND, 1948.

Braun, Wernher von. "German Rocketry," *The Coming of the Space Age*. Edited by Arthur C. Clarke, New York: Meredith Press, 1967.
Braun, Wernher von. "Major U. S. Programs," *The History of Rocket Technology*. Edited by Eugene M. Emme, Detroit: Wayne State University Press, 1964.
Braun, Wernher von, Ordway, Frederick I., and Dooling, David. *Space Travel: A History*. New York: Harper and Row Publishers, 1985 (4th edition of a book previously published under the title *History of Rocketry and Space Travel*.)

Caidin, Martin, "Vergulstungwaffe 2," *Mechanix Illustrated*. March, 1945.

Cooksley, Peter G. *Flying Bomb*. New York: Charles Scribner & Sons, 1979.

Dawson, P. J. *German Organisation and Personalities Engaged in Research and Development of Armaments During the Second World War*. Ministry of Supply, London: Report No. 436/I, 1948.

Department of the Army. *Army Ordnance Department Guided Missiles Program*. 1 January 1948.

Dornberger, Walter. "The Lessons of Peenemünde." Technical Intelligence Branch, OWC, GU-15, 460, N. D., (ca. 1958).
Dornberger, Walter. *V-2*. New York: Viking Press, 1954.
Dornberger, Walter. "The German V-2," *The History of Rocket Technology*. Edited by Eugene M. Emme, Detroit: Wayne State University Press, 1964.

Durant, Frederick C., III. "Robert H. Goddard and the Smithsonian Institution," *First Steps Towards Space*. Washington, D. C.: Smithsonian Institution Press, 1974.

Emme, Eugene M. *Aeronautics and Astronautics, An American Chronology of Science and Technology in the Exploration of Space 1915 – 1960*. Washington, D. C.: U.S. Government Printing Office, 1961.

Englemann, Joachim. *V2 Dawn of the Rocket Age*. West Chester, PA: Schiffer Publishing, 1990.

Eyestone, S. F. *A Study of the German Type II Accelerometer*. Report #AL-123, Los Angeles, North American Aviation, Inc., 1947.

Fedden, Roy. *The Fedden Mission to Germany*. Combined Intelligence Objectives Subcommittee, London: 1945.

Felkin, S. D. *The A-4 Rocket – Further Information*. A.D.I. (K) Report No. 228/1945, March 1945.
Felkin, S. D. *More Information on the A-4 Rocket*. A.D.I. (K) Report No. 34B/1945, January 7, 1945.

Fogel, H. M. *Fuzing System of the German A-4 Rocket (V-2)*. Combined Intelligence Objectives Subcommittee, G-2 Division, SHAEF, Items No. 3 and 4, File No. XXVII-37, April 1945.

Bibliography

Fraser, L. W., and Siegler, E. H. *High Altitude Research Using the V-2 Rocket, March 1946 – April 1947*. Bumble Series Report #81, Silver Spring, Maryland: The Johns Hopkins University Applied Physics Laboratory, July 1948.

Friedman, Henry. *Summary Report on A-4 Control and Stability*. Wright Field, Dayton, Ohio: Headquarters Air Materiel Command, Report #F-SU-2152-ND, 1947.

Garlinski, Josef. *Hitler's Last Weapons*. New York: New York Times Books, 1978.

Garstens, M. A., Newell, H. E., and Siry, J. W. *Upper Atmosphere Research Report Number 1*. Naval Research Laboratory Report No. R-2955, Washington, D. C.: Office of Naval Research, October 1, 1946.

Gatland, Kenneth. *Missiles and Rockets*. New York: MacMillan Publishing Company, 1975.

German Long-Range Rocket Projectile – Storage and Launching Sites. A. I. 2(g) Report No. 4/x, August 9, 1944.

"German Propaganda and the Rocket", *Air Ministry Weekly Intelligence Summary Number 281*, 20 January 1945.

Hanrahan, James S., and Bushnell, David. *Space Biology*. New York: Basic Books, Inc., 1960.

Harford, James. *Korolev*. New York: John Wiley & Sons, Inc., 1997.

Hargast, W. J. "Assembly Technique on V-2 Bombs Revealed in Underground Factory," *American Machinist*. August 16, 1945.

Helfers, M. C. *The Employment of V-Weapons by the Germans During World War II*. Washington, D. C.: Office of the Chief of Military History, U. S. Army, May 31, 1954.

Henry, James P., et. al. "Animal Studies of the Subgravity State During Rocket Flight," *Journal of Aviation Medicine*. V. 23, No. 10, October 1952.

Hintze, Guenther. *Special Report Missile I*. Fort Bliss, Texas: Department of the Army Ordnance Research & Development Division Suboffice (Rocket), June, 1949.

Historical Division, Air Force Missile Development Center. *History of Research in Space Biology and Biodynamics at the Air Force Missile Development Center, Holloman Air Force Base, New Mexico, 1946 – 1958*. Holloman Air Force Base, New Mexico: 1958.

Hüzel, Dieter. *Peenemünde to Canaveral*. Englewood Heights, New Jersey: Prentice-Hall, Inc., 1962.
Interim Report on Large Sites, Rocket Firing Platforms and Rocket Storage Sites in the Pas de Calais Area, A. I. 2. (L) Report No. 114, October 7, 1944.

Irving, David. *The Mare's Nest*. Boston: Little Brown and Company, 1965.

Kay, Anthony L. *Buzz Bomb*. Monogram Close-Up 4, Boylston, Massachusetts: Monogram Aviation Publications, 1977.

Kennedy, Gregory P. *Vengeance Weapon 2: The V-2 Guided Missile*. Washington, D. C.: Smithsonian Institution Press, 1983.

Klee, Ernst, and Merk, Otto. *The Birth of the Missile*. New York: E. P. Dutton & Co., 1965.

Kooy, J. M. J., and Uytenbogaart, J. W. H. *Ballistics of the Future*. Haarlem, Holland: The Technical Publishing Company H. Stam, 1946.

Lasby, Clarence G. *Project Paperclip*. New York: Ateneum, 1971.

Lehman, Milton. *This High Man*. New York: Farrar, Straus and Company, 1963.

Ley, Willy. *Rockets, Missiles, and Men in Space*. New York: The Viking Press, 1968.

Magazine & Book Section, Public Information, Navy Department. "Navy Fires V-2 Rocket From Deck of USS Midway." Washington, D. C., Navy Department, September 8, 1947.

Mallan, Lloyd. *Men, Rockets and Space Rats*. New York: Julian Messner, Inc., 1955.

Martenson, C. D. "Operations Backfire and Clitterhouse (British Firings of V-2 Rockets)". London: Military Attache, Report R 5499-45, 1945.

Michel, Jean. *Dora*. New York: Holt, Rinehart, and Winston, 1979.

Middlebrook, Martin. *The Peenemünde Raid*. London: Allen Lane, 1988.

Mossop, I. A., *The Electrolytic Integrating Accelerometer for the Automatic Control of Range of the German A-4 Rocket*. Farnborough Hants: Royal Aeronautical Establishment, R.A.E. Report EL.1387, 1946.

Munson, Kenneth G., *Aircraft of World War Two*. London: Ian Allen Ltd., 1962.

Nebel, Rudolph, "Rocket Flight to the Moon – From Idea to Reality: A Memoir," *Essays on the History of Rocketry and Astronautics: Proceedings of the Third Through the Sixth History Symposia of the International Academy of Astronautics, Volume II*. Edited by R. Cargill Hall, Washington, D. C.: National Aeronautics and Space Administration, NASA Conference Publication 2014, 1977.

Newell, H. E., and Siry, J. W., *Upper Atmosphere Research Report Number II*. Naval Research Laboratory Report No. R-3030, Washington, D. C.: Office of Naval Research, December 30, 1946.

New York Times, "V-2 Assembly Plant is Found in Mountain," April 14, 1945.

Oberkommando des Heeres (High Command of the Army). *Das Gerät A-4, Baureihe B*, (The Device A-4, Model B). January 2, 1945 (translated by the General Electric Company, Schenectady, New York.)

Oberth, Hermann. "My Contributions to Astronautics," *First Steps Towards Space*. Washington, D. C.: Smithsonian Institution Press, 1974.

Ordway, Frederick I., and Sharpe, Mitchell R. *The Rocket Team*. New York: Thomas Y. Crowell Publishers, 1979.

Raushenbach, B. V., and Biryukov, Yu. V., "S. P. Korolyev and the Development of Rocket Engineering," *First Steps Towards Space*. Washington, D. C.: Smithsonian Insititution Press, 1974.

Rhea, John, ed., Berlin, Peter, trans. *Roads to Space*. New York: McGraw-Hill Companies, 1995.

Ross, Harry E. "The British Interplanetary Society's Astronautical Studies, 1937 - 39," *First Steps Towards Space*. Washington, D. C.: Smithsonian Institution Press, 1974.

Santiago, Dawn Moore, ed. *New Mexico Space Journal, Number 1*. Alamogordo, New Mexico: New Mexico Museum of Space History, June 2001.

Schultz, H. A. *Technical Data on the Development of the A-4 (V-2)*. Huntsville, Alabama: The George C. Marshall Space Flight Center, 1965.

Simons, David G. *Use of V-2 Rockets to Convey Primates to Upper Atmosphere*. Air Force Technical Report 5821, Dayton, Ohio: Air Force Materiel Command, May 1959.

Speer, Albert. *Inside the Third Reich*. New York: MacMillan Publishing Company, 1970.
Speer, Albert. *Infiltration*. New York: MacMillan Publishing Company, 1981.

Storage and launching of German Long-Range A-4 Rocket Projectile. A. I. 2(g) Report No. 6/x, August 25, 1944.

Bibliography

Thomas, Shirley. *Men of Space, Volume 1*. Philadelphia: Chilton Book Company, 1960.

United States Army Ordnance Corps and the General Electric Company. *Hermes Guided Missile Research and Development Project 1944 – 1954*. Technical Liaison Branch, Chief of Army Ordnance, September 25, 1959.

United States Strategic Bombing Survey, Aircraft Division Industry Report. *Strategic Bombing of the German Aircraft Industry, Chapter IX, Report on V-Weapon Production*. 1945.

United States Strategic Bombing Survey, Military Analysis Division, *V-Weapons (Crossbow) Campaign*, January 1947.

United States Strategic Bombing Survey, Physical Damage Division. *V-Weapons in London*. January 1947.

United States War Department. *Handbook on Guided Missiles of Germany and Japan*. Washington, D. C.: 1946.

War Office. *Report on Operation Backfire*. London: Ministry of Supply, 1946.

White, L. D, *Final Report, Project Hermes V-2 Missile Program*. Schenectady, New York: General Electric, Report No. R52A0510, September 1952.

Wilke, Hugh H., and Steiner, Walter A. *Survey of Graphite Rudders for V-2 Rockets*. Joint Intelligence Objectives Agency, Washington, D. C.: FIAT Final Report No. 105, 1945.

Winter, Frank H. *Prelude to the Space Age: The Rocket Societies: 1924 – 1940*. Washington, D. C.: Smithsonian Institution Press, 1983.

Index

A-1, 12, 60
A-2, 12, 60, 126
A-3, 12 – 15, 13, 16, 18, 126
A-4, 14, 16, 19, 20 – 22, 24 – 25, 26, 30 – 34, 38, 39 – 41, 42, 43, 47, 62
A-4b, 125
A-5, 16, 18
A-9/A-10, 125
Alamogordo, New Mexico, 94, 95
Albert I, 104 – 105
Albert II, 105
Albert III, 105
Albert IV, 105

Backfire, Operation, 84 – 87
Blizna, Poland, see Heidelager
Braun Werhner von, 10 – 13, 16, 20, 26, 29 – 30, 42, 83, 114 – 116, 118, 126
British Interplanetary Society, 86 – 87
Bumper Program, 106 – 109, 122, 124, 127

Crow, Alwyn (Sir), 26, 84

Degenkolb, Gerhard, 20, 32
Dora Concentration Camp, 32, 34
Dornberger, Walter, 8, 11, 12, 13, 14, 16, 18 – 21, 25 – 26, 29 – 32, 39, 42 – 43, 47, 83, 84, 125, 126

Fiesler – 103 (V-1), 23 – 25, 44, 64 – 65, 79

Glushko, Valentin, 89, 90
Goddard, Robert Hutchings, 6 – 7, 45, 92, 126

Heidelager, 31, 41, 44, 47, 89
Henry, James P., 104 – 105
Hermes II, 109 – 111, 123 – 124
Himmler, Heinrich, 21 – 22, 30, 32, 42, 47
Hitler, Adolf, 18, 20, 26, 30, 47
Hydra, Operation, 28 – 29

Jupiter-C, 115, 116, 118

Kammler, Hans, 32, 42, 63 – 66, 83, 126
Kapustin Yar, 90, 127
Korolev, Sergei Pavlovich, 87 – 91, 115 – 116, 118
Kummersdorf (Army Experimental Station West), 8, 11, 12, 13, 14, 16, 18

Malina, Frank, 7, 93
Marshall Space Flight Center, 116, 118
Midway, USS, 102 – 103
Mirak, 9, 10, 11, 12

Nebel, Rudolph, 9, 11, 12

Oberth, Hermann, 6, 7, 8 – 9, 10

Paperclip, Operation, 92, 94, 111
Peenemünde, 13, 14, 16 – 22, 24 – 25, 26 – 31, 44, 47, 83, 126
Pushover, Operation, 103

R-1, 90, 91, 115, 127
R-2, 91, 115
Raketenflugplatz, 10 – 11, 13
Redstone, 113, 114 – 116, 118, 127
Reitsch, Hanna, 28
Rickhey, Georg, 36, 80 – 81
Sandy, Operation, 102 – 103, 122
Saturn I, 118, 127
Saturn IB, 118, 119, 127
Saturn V, 118, 120, 121, 127
Schmidt, Paul, 23
Simons, David G., 104 - 105
Speer, Albert, 19, 20, 25, 42, 47

Thiel, Walter, 14 – 16, 18, 29
Tsiolkovsky, Konstantine, Eduardovich, 6, 89
Turner, Harold R., 98, 105, 106

V-2, 44, 48 – 62, 63 – 66, 69, 79, 80 – 86, 89, 92 – 112, 121, 122 – 124, 126 - 127

WAC Corporal, 92, 93, 95, 99, 106 – 109, 126 – 127
White Sands Proving Ground, 92 – 112, 122, 123, 126 - 127
Winkler, Johannes, 9, 10, 126

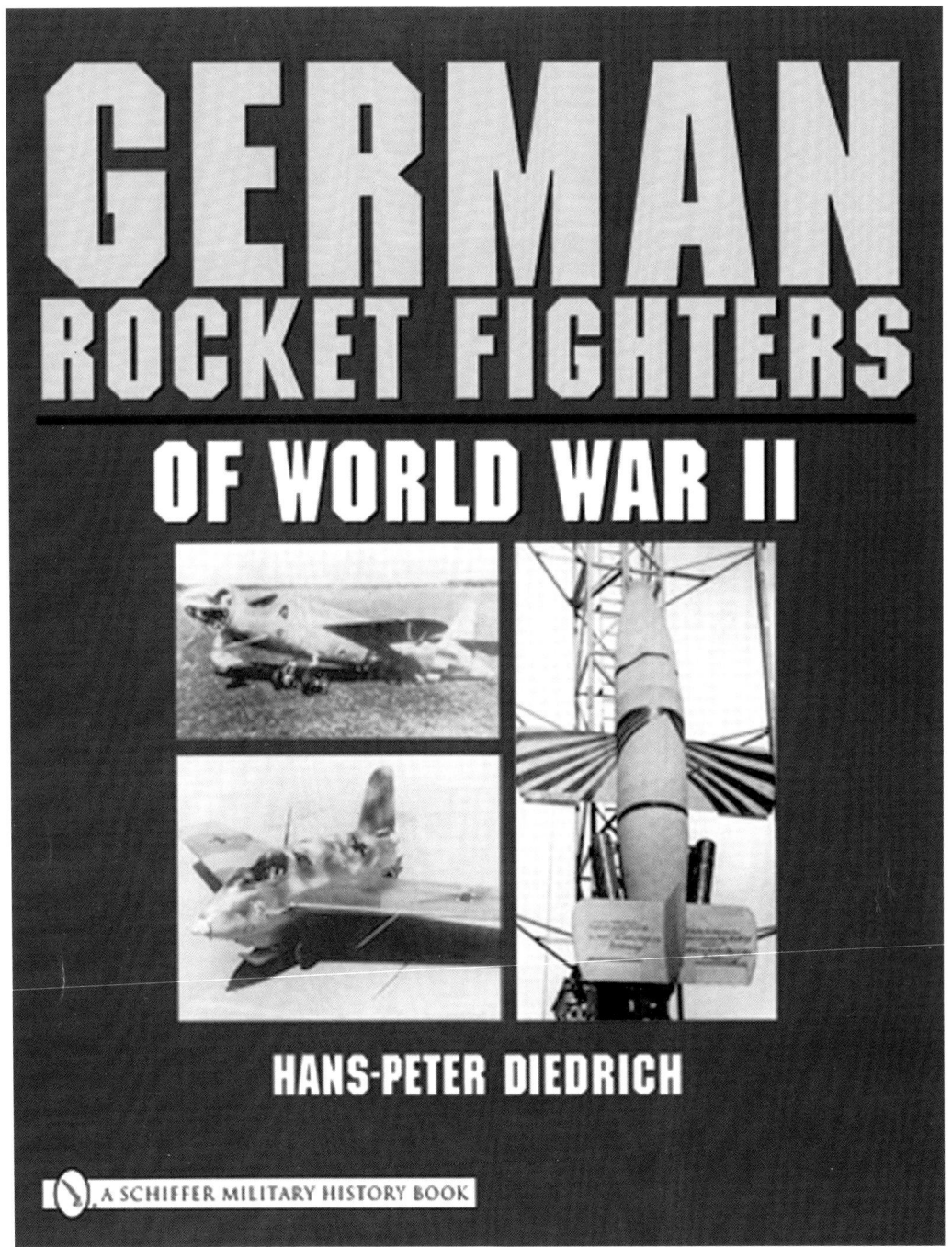

German Rocket Fighters of World War II
Hans-Peter Diedrich

This book is the first to present every rocket aircraft flown in Germany and the rocket systems developed by Walter and BMW, as well as several of the most interesting projects drawn up by Germany's aviation industry. In 1940 the Deutsche Forschungsanstalt für Segelflug launched the DFS 194, developed by Alexander Lippisch, and the tests with this experimental plane laid the foundation for the Messerschmitt Me 163, the world's first operational rocket fighter. The technology incorporated into Germany's rocket planes – the Messerschmitt Me 163 Bs and Cs, the Me 263, and Bachem Ba 349 Natter – was recognized throughout the world as cutting edge and after the war had a major impact on the technological development of other countries. This book is a must-have for every aviation enthusiast.

Size: 81/2"x11" ■ over 90 b/w photographs, line schemes, charts ■ 152 pp
ISBN: 0-7643-2220-6 ■ hard cover ■ $49.95